TOWARD ZERO CARBON

The Chicago Central Area DeCarbonization Plan

TOWARD ZERO CARBON
The Chicago Central Area DeCarbonization Plan

ADRIAN SMITH + GORDON GILL
ARCHITECTURE

images
Publishing

Published in Australia in 2011 by
The Images Publishing Group Pty Ltd
ABN 89 059 734 431
6 Bastow Place, Mulgrave, Victoria 3170, Australia
Tel: +61 3 9561 5544 Fax: +61 3 9561 4860
books@imagespublishing.com
www.imagespublishing.com

The Images Publishing Group Reference Number: 952

National Library of Australia Cataloguing-in-Publication entry:

Title:	Toward zero carbon : the Chicago Central Area decarbonization plan / Adrian Smith + Gordon Gill Architecture.
ISBN:	9781864704334 (pbk.)
Subjects:	Carbon—Environmental aspects—Illinois—Chicago. Carbon dioxide mitigation—Illinois—Chicago. Pollution prevention—Illinois—Chicago. Chicago (Ill.)—Environmental conditions.

Other Authors/Contributors: Adrian Smith + Gordon Gill Architecture.

Dewey Number: 363.738740977311

Edited by Kevin Nance and Debbie Fry

Pre-publishing services by Mission Productions Limited, Hong Kong

Printed on FSC-certified 157 gsm Sun FSC Matt paper by Everbest Printing Co. Ltd., in Hong Kong/China

A NOTE ABOUT THIS BOOK

This is a book about Chicago. It's about the future of our home city as the greenest in America and a world leader in energy efficiency and reducing carbon emissions. In these pages, we offer a comprehensive vision of how Chicago can become much more sustainable, in part by upgrading and in some cases re-purposing existing buildings, and how we can significantly limit the city's contribution to the causes of climate change. We also envision a series of innovations in urban design in which a new picture of downtown Chicago emerges. In it, families live, work, send their children to school, go to the grocery store and stroll landscaped promenades within a dense urban core, all without needing a car. Quality of life is high, carbon footprints low. If Chicago is the city that works, this future Chicago is the city that works even better.

Chicago Loop, 2010

But this book is potentially about other cities, too. Properly adapted to local conditions, the Chicago Central Area DeCarbonization Plan could form a foundation for custom-designed plans for Atlanta or Los Angeles, Rome or Rotterdam, Bangkok or Beijing. That's because cities everywhere are facing many of the same problems confronting Chicago. The highest percentage of commercial structures dates from the mid-20th century, an era in which energy was relatively cheap and there was limited concern about wasting it. Traffic congestion and air pollution are nearly universal urban problems; some of the largest metropolises are approaching a state of perpetual gridlock. In most of these cities, the percentage of residential buildings in the urban core remains remarkably low, even as older, obsolete office buildings, which could be converted to apartments and condominiums, sit largely empty. At the same time, families avoid living downtown because of the lack of amenities—particularly schools and green space—necessary for raising children. And so the pattern of long, stressful commutes in machines powered by fossil fuels continues, dumping untold metric tons of carbon into the atmosphere. But it's a pattern that can be broken, which is what *Toward Zero Carbon* is about.

As we continue to develop the DeCarbonization Plan for Chicago, then, we also look to apply its lessons elsewhere. This year we launched PositivEnergy Practice (www.pepractice.com), a consulting firm whose charter includes designing site-specific DeCarbonization Plans for cities around the world. As the saying goes, think locally, act globally. We plan to.

Adrian Smith, Gordon Gill and Robert Forest
October 2010

FOREWORD: THE ART AND SCIENCE OF CITY-MAKING

Ned Cramer, Editor-in-Chief, *Architect* magazine

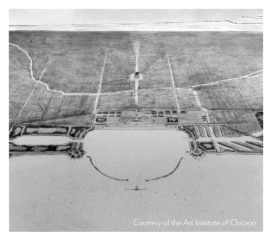

Courtesy of the Art Institute of Chicago

Jules Guérin rendering for the *Plan of Chicago*, 1909

The comparison is inevitable. By creating a green plan for Chicago's downtown Loop, Adrian Smith and Gordon Gill are standing on the shoulders of giants. For no architect or planner can think, much less build, on an urban scale in Chicago without taking into account the 1909 *Plan of Chicago* by Daniel H. Burnham and Edward H. Bennett. Jules Guérin's famous watercolor renderings of the *Plan*, housed in the Art Institute of Chicago, depict an orderly scheme of streets, parks and waterways, with monumental buildings aligning on precise Beaux-Arts axes. Beneath the surface of these beautiful images, however, lies Burnham and Bennett's altruistic desire to improve the city and the lives of its citizens. The civilizing, classical order of the plan would serve as an antidote to the industrial blight and workers' plight that Upton Sinclair had exposed in his 1906 novel *The Jungle*. Chicago would supplant its industrial origins to become a Paris on the Prairie.

The salutary effects of the 1909 *Plan* are discernable throughout Chicago: the city's transportation network, its 26-mile public lakefront, its necklace of landscaped boulevards and parks, even its official motto, *urbs in horto*, "City in a Garden." The stench of the city's meatpacking plants and the soot of its coal-powered factories dissipated long ago, but environmental dangers remain, largely in the CO_2-producing form of vehicles and buildings. Enter Smith and Gill, the Burnham and Bennett of our age, equally intent on cleaning up Chicago through design with their Chicago Central Area DeCarbonization Plan.

Many contemporary architects strive to design and build sustainably, with the least possible negative effect on the environment. But the profession as a whole still struggles to expand its role beyond the act of making big, beautiful objects. For more than a decade, spectacular projects such as Frank Gehry's Guggenheim Museum Bilbao have dominated architectural discourse. Star architects and blockbuster buildings keep architecture in the public eye, which benefits the profession, but along with the accolades comes the implication that architecture is a luxury rather than a staple. The DeCarbonization Plan, by contrast, identifies a building's performance as the critical design imperative; design remains a core value, but the definition of design has evolved to encompass more than aesthetics. What Smith and Gill offer on an urban scale is not simply a collection of shapely buildings; it's an integrated system for the management of the city's resources, predicated upon scientific analysis.

Compared to the 1909 *Plan*, then, there is relatively little architecture in the DeCarbonization Plan. Smith and Gill make the case for an improved city not through pictures of fantastical new structures and cityscapes, à la Guérin, but through charts and tables that survey carbon emissions and other information about existing buildings and new strategies for the distribution of people, energy and waste. To be sure, there are some exciting, symbolically resonant design proposals here, such as the Chicago Eco-Bridge that Smith and Gill propose in Lake Michigan, in line with Burnham's Grant Park. But the plan's real emphasis is on data.

Burnham once said, "Make no little plans." Smith and Gill make lots of little plans, minor tweaks and adjustments, that could have major cumulative effects. The architects are not designing a *tabula rasa* new Chicago, assuming dictatorial powers to tear down and rebuild at will. Instead, they embrace current realities and adapt the existing city through economical and often very pragmatic means. The DeCarbonization Plan includes a section entitled "Pneumatic Waste Collection With Plasma Arc Gasification and Aggressive Recycling Program." Ten years ago, it would have seemed unthinkable for an architect of the stature of Adrian Smith, a designer of trophy skyscrapers such as Dubai's Burj Khalifa and Chicago's Trump Tower, to concern himself (or herself) with the minutiae of waste management. This change exemplifies an industry-wide redefinition of an architect's core expertise.

In 2006, Smith left a decades-old perch at the blue-chip firm Skidmore, Owings & Merrill, and entered into a new partnership with Gill, a younger SOM alum. Since that time, Smith has entered a new phase in his career as a master builder, and Gill has emerged at his side as one of the most prominent members of an emerging generation of architects. While a seemingly unlikely pair, together they stand at the forefront of a sea change in the process, practice and product of architecture—a change that supplements the architect's traditional creative instincts with more sober, scientific methods. A sure sign of all this is the 2010 R+D Award for architectural technology that the DeCarbonization Plan received from *Architect*.

In one grim and entirely possible future scenario, a population accustomed to the advantages of cheap energy must struggle to make do with diminishing resources. The DeCarbonization Plan offers a welcome alternative. Architects are increasingly being held accountable for the performance of the buildings they design. Smith and Gill don't grudgingly accept this greater responsibility; they embrace it. The Chicago Central Area DeCarbonization Plan demonstrates that through careful stewardship, scientific thinking and no small amount of creativity, the future may not be a struggle at all.

Chicago Eco-Bridge

A VISION FOR CHICAGO

Willis (formerly Sears) Tower with proposed hotel, Chicago

Chicago has never refrained from making big plans. We're a city known for innovative solutions to complex problems.

In 2007, Adrian Smith + Gordon Gill Architecture was hired by the owners of Willis (then Sears) Tower to design a sustainable renovation, modernize the landmark building and improve energy savings and efficiency. We conducted a comprehensive study and developed a series of improvements that would reduce the base building electricity use by 80 percent (including energy savings and co-generation), equivalent to 68 million kilowatt hours or 150,000 barrels of oil per year. At the same time, we designed a highly sustainable, 50-story five-star hotel to be built on the Willis Tower plaza. The hotel will operate using part of the energy savings from the Willis Tower greening project, meaning it will draw net zero energy from the power grid. By combining energy-efficient design with the environmental and social sustainability inherent in a mixed-use development, these projects will transform the block into a high-performance district.

In the process, we realized that the implications of the Willis Tower transformation were applicable to the entire downtown Loop. New, high-performance development needed to go hand-in-hand with the modernization of the existing building stock; they should complement each other in terms of both program and energy use. Mixed-use, walkable neighborhoods were needed, which called for a new approach to urban design. This holistic view then led us to consider several other elements that feed into the downtown grid: waste systems, water systems, energy networks, transportation and infrastructure. Community engagement and funding strategies were also designed. The result is a bold and comprehensive DeCarbonization Plan that recognizes and embraces the fact that in our pursuit of a sustainable future for Chicago, everything is connected.

Yes, the DeCarbonization Plan is as complex as the matrix of issues it addresses, and its implementation will be a major undertaking. But we believe that it represents a viable way forward as the city pursues its carbon reduction goals for 2020, 2030 and beyond. In any case, Chicago is a city that has always loved a challenge.

INTRODUCTION

What if we could tap into Chicago's latent potential by using the existing built environment as a carbon asset? What if we could redefine energy as a commodity to be traded between buildings, blocks, cities? What if we could transform Chicago's Loop into a net carbon-positive district?

Today, the world faces an unprecedented environmental crisis. We must begin to reverse the damage inflicted on the planet since the Industrial Revolution. In 2002, the 2030 Challenge stated, "scientists give us 10 years to be well on our way toward global greenhouse gas emissions reductions in order to avoid catastrophic climate change." In 2008, Chicago formulated the Chicago Climate Action Plan to begin to address these issues. Today, we know that moving forward in a sustainable manner requires more than simply building new energy-efficient buildings or increasing the number of hybrid cars on our roads. We must make significant, demonstrable changes to our existing city landscape, altering not only how our city looks but how it works. An urban ecosystem relies on the true integration of each of a city's elements. Smart buildings rely on smart transit networks; smart energy systems rely on the creation of smart infrastructure. The process of DeCarbonization aims to improve the performance of every major metropolitan system to create a healthier, more sustainable, more livable city.

The Chicago Central Area DeCarbonization Plan proposes a philosophical shift. It aims to simultaneously reduce the carbon emissions of downtown Chicago while creating a vibrant urban environment. Unlike other plans, this is not a checklist. It's a holistic approach to augmenting the environmental, economic and qualitative aspects of Chicago city life. The Plan distinguishes itself from the plans of other cities by focusing on an integrated approach to solving problems. The study examines all of the carbon sources characteristic of the urban condition, but does not simply reach a reduction number calculated from an applied set of assumptions. The plan sets up a framework for maintaining the economic and cultural vitality of the urban core from an energy and carbon perspective. The continued viability of cities and urban living depends upon the idea that growth can continue without a negative environmental impact. DeCarbonization allows cities to develop as centers of healthy, multifaceted lifestyles with minimal environmental footprints.

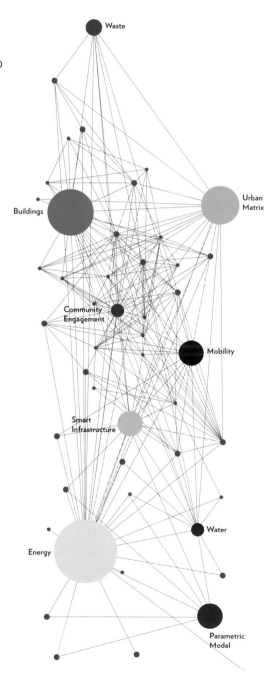

Representation of holistic approach required to identify and find solutions to carbon emissions

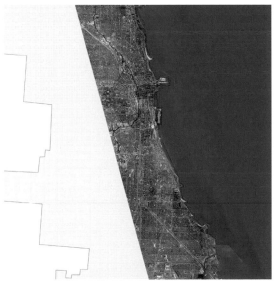

The Loop is the central business district in Chicago's downtown.

On the surface, there are simpler ways to reduce carbon. A wind farm or large-scale photovoltaic array can power hundreds of buildings. A single nuclear plant can power an entire region. Why, then, is there a need for a DeCarbonization Plan for the Loop?

If we could simply replace greenhouse-gas-emitting fossil fuel plants with renewable energy resources, urban DeCarbonization would be unnecessary from an energy perspective. But a strong DeCarbonization Plan addresses more than energy use and carbon emissions; it also addresses the state of existing building stock. It creates an opportunity for a true paradigm shift in the way citizens engage with their cities. Creation of new parks and public spaces, improvements to both the existing infrastructure and the integration of new methods of transit, foundation of community action groups and the education of our younger citizens are all critical components of a strong DeCarbonization Plan. Without continual improvement to these aspects of the urban environment, cities will decay over time.

Solutions to critical city issues must also address the complexities of carbon production. Chicago's current building stock consumes much more energy than necessary, largely due to the age of the majority of downtown structures. It does not make economic or ecological sense to demolish and replace the real estate that has been developed over the last 40 years. Upgrading these buildings will both reduce carbon emissions and ensure future economic viability. Beyond efficiency savings for owners and tenants, changing the building stock to be truly sustainable creates an opportunity for a

shift in the economic culture of real estate. By their transformation into high-performance structures, aging buildings increase in value and tap into the potential to transfer excess energy loads back to the grid—all while offsetting the need for new construction.

In addition to the "Buildings" chapter, the Chicago Central Area DeCarbonization Plan analyzes seven areas of carbon production and reduction relevant to the urban condition: "Urban Matrix," an analysis of real estate and land use; "Smart Infrastructure," an exploration of systems designed to optimize resource performance; "Mobility," an assessment of transit and connectivity; "Water," an evaluation of how this critical resource is used and conserved; "Waste," an assessment of citywide processes and systems for reducing, recycling and disposal; "Community Engagement," the development of programs to engage our citizens; and "Energy," an examination of existing and new energy sources.

By addressing these issues not independently but as integrated systems, the Chicago Central Area DeCarbonization Plan advances beyond offering only technological fixes, identifying innovative methods for improvement suited both to the nature of the climate change problem and the creation of a strong urban environment. Chicago will illustrate to the international community that urban quality of life can continue to improve, despite—and in response to—the global issue at hand. Chicago will emerge as a global leader, epitomizing the sustainable city and continuing to be a center for architectural and urban innovation.

BUILDINGS

In the United States, buildings account for an average of 40% of carbon emissions. In Chicago, that number soars to 70%. This disparity is caused by the fact that our transportation footprint is relatively small. But as in many cities of similar age, many of Chicago's buildings feature inefficient and outdated systems that cause high energy loads.

This chapter analyzes Chicago's existing building carbon emissions and energy usage by considering the components that are responsible for the majority of loads and by looking closely at how the different building types in the Loop perform.

We outline procedures to transform the existing Chicago Loop building stock into a more efficient neighborhood comprised of buildings that enjoy symbiotic relationships and produce energy.

URBAN MATRIX

Commercial space in Chicago's downtown Loop accounts for 90% of its land use and 97% of its carbon emissions. By contrast, the Loop contains almost no residential space and virtually none of the amenities and support infrastructure that many homeowners value in their neighborhoods, such as schools, daycares, parks or grocery stores. This virtually forces those who work in the area to look elsewhere in the city for suitable housing and living environments. Analysis also confirms the lower carbon footprint created by mixed-use, 24-7 high-density communities.

This chapter examines a number of approaches to creating a vibrant urban core in the downtown area, including increased amenities and new schools. Design solutions include the conversion of select existing buildings into new residential buildings, the addition of new high-performance commercial towers, a new parkway along Monroe Street, a larger presence of green roofs, new pocket parks and a new Daley Environmental Learning School.

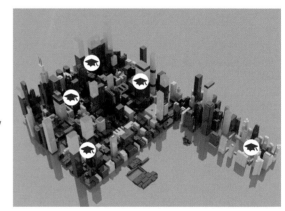

MOBILITY

To successfully entice more of the population to use sustainable modes of transit, the city needs to augment existing transit and incorporate new systems to accommodate commuter needs.

This chapter analyzes the relationship between urban density and gas consumption. Cities that have denser cores generally have lower per-capita gasoline consumption. A further analysis looks at rail and bus commuting, bicycle and pedestrian routes and the existing public service fleets, including taxis.

Logical, integrated solutions, as well as new trends, are explored. Design solutions are proposed to enhance the quality of life of Loop residents and commercial tenants.

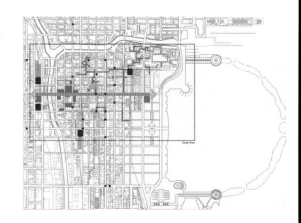

SMART INFRASTRUCTURE

Energy consumption in the United States (and, correspondingly, in the city of Chicago) has dramatically escalated over the past several decades, significantly outpacing population growth as new devices have become ubiquitous in our daily lives. While new technologies have led to increased strain on an aging infrastructure, they also present new opportunities to improve the intelligence and distribution efficiency of energy and information, engendering new infrastructure intelligence. This chapter examines regional trends in energy supply and distribution and focuses on low-carbon technologies. Strategies such as smart grids are expanded to encompass the broader scope of smart infrastructure, which goes beyond supply-and-demand management strategies and seeks to add city amenities through information technologies.

We consider four distinct networks: the real estate, public realm, transportation and utilities sectors. We identify specific Chicago projects, such as a clean energy taxi program and a vehicle-to-grid car-sharing network.

WATER

It's difficult for many Chicagoans to imagine that the high-quality water they enjoy might be a source of greenhouse gas emissions. To a degree, this mindset is understandable, given the proximity and relatively pure nature of our main water source. But in fact, the purification, delivery and heating of potable water, as well as the treatment of waste water, can be significantly improved to reduce the overall carbon footprint of water in Chicago.

This chapter studies two broad areas to investigate methods for carbon reduction. Reducing the carbon load in a gallon of water is followed by a look at conservation efforts to create a cascading strategy for carbon reduction.

Design solutions present several possibilites for the carbon loading and conservation sides of the carbon reduction effort for water. Solutions range in scale from the replacement of aerators on sinks to the restoration of wetlands.

WASTE

A large percentage of the municipal waste generated in Chicago can be recycled or reused. But to achieve the reduction goals of the future, the city needs to augment the existing waste infrastructure. In the spirit of conservation, Chicago can reuse existing systems, such as the coal tunnels, in new ways to efficiently collect and transport waste. Much of what can't be recycled can be converted into clean new forms of energy, reducing carbon even further.

The analysis section identifies waste materials that the city should focus upon. Existing and new waste initiatives, as well as the use of an existing tunnel framework to collect and channel waste out of the city, will help meet Chicago's current carbon goals.

Design solutions offer possibilities for reduce the carbon load of waste treatment, such as repurposing the city's existing network of underground tunnels as a pneumatic waste removal system.

COMMUNITY ENGAGEMENT

Participation in community activities enhances shared feelings of citizenship and pride. The expansion of social networks from new technologies will strengthen and change both identification and interaction between fellow Chicagoans. While citizens can use these new social mediums to create larger communities dedicated to sustainability, they can also refocus their energy to develop action-oriented groups that provide incentives for more sustainable attitudes.

The strategies within this chapter examine concepts of branding, green team organizations, multilingual advertising and educational materials and social marketing programs. By promoting change, Chicagoans will become more aware of and ultimately involved in the carbon reduction initiative. A series of proposed initiatives, from urban agriculture to large-scale learning environments, will inspire and educate the community to incorporate these ideals into their everyday lives.

ENERGY

Rapid growth in energy demand in the United States is taxing an aging energy infrastructure, leading to brown-outs and lost economic output. Moreover, it has led to increased reliance on foreign energy sources, creating the potential for geopolitical tensions and consequential environmental damage.

In this chapter, regional trends in energy supply for the state of Illinois and Chicago are studied in aggregate and broken down by energy source (coal, natural gas, nuclear and hydroelectric). Trends are projected for 2020, taking into account pending federal legislation that would mandate reductions in aggregate carbon emission factors, suggesting an increased reliance on low-carbon technologies such as wind and photovoltaic energy. These systems are also compared based upon their capital intensity and annual carbon abatement costs.

We also examine strategies for on-site renewable and co-generation of energy, complemented by off-site renewable and district energy strategies. We propose three distinct scales of solutions that integrate elements of a new energy equation within Chicago's built environment.

© iStock/Kristy Pargeter

PARAMETRIC MODEL

The research for the DeCarbonization Plan acted as a catalyst for the development of a design and planning tool to calculate the carbon savings achieved by various changes made to the urban environment. This parametric model will function as a part of the data model, though where the data model focused on the existing condition of the city, the parametric model is concerned with the future of the city and the potential impact on performance and sustainability that discerning change will enable. As in the data model, the power of this tool is in its potential to assess the city as a functioning whole, with the ability to change any single parameter and to understand how local, bottom-up change can have great effect on the whole.

FUNDING

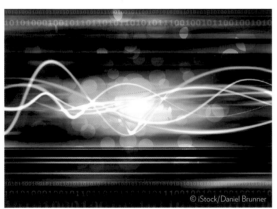

© iStock/Daniel Brunner

The DeCarbonization Plan has the potential to tap a wide array of funding resources. In this chapter, several federal, state, municipal and private-sector resources are described and discussed in terms of which would work best for funding a successful DeCarbonization program in the Loop. There are a number of properties and resources in the study area to achieve these goals. A key consideration will be aligning the resources so that they reinforce one another; so too with the properties. Our overarching goal will be an integration of a number of the resources and the participation of several properties so that a district is formed wherein many owners undertake significant energy improvements. At least two things are paramount: (1) ensuring that major game-changing DeCarbonization projects are launched and (2) creating a groundswell, so that many DeCarbonization initiatives are undertaken by a wide group of owners in the wake of these pilot projects.

CHICAGO CLIMATE ACTION PLAN

To do its part to avoid the worst global impacts of climate change, Chicago needs to significantly reduce carbon emissions. As part of the Chicago Climate Action Plan (CCAP), an initial goal of a 25% reduction below 1990 levels for 2020 was proposed for new and significantly renovated buildings. The 2030 Challenge has a more aggressive 80% reduction goal for 2020 in carbon emissions for new and renovated buildings.

OUR CITY. OUR FUTURE.

© Chicago Climate Action Plan

Strategies of the Chicago Climate Action Plan (CCAP):

Energy-efficient buildings
- Retrofit commercial and industrial buildings
- Retrofit residential buildings
- Trade in appliances
- Conserve water
- Update city energy code
- Establish new guidelines for renovations
- Cool with trees and green roofs
- Take easy steps

Clean and renewable energy sources
- Upgrade power plants
- Improve power plant efficiency
- Build renewable electricity
- Increase distributed generation
- Promote household renewable power

Improved transportation options
- Invest more in transit
- Expand transit incentives
- Promote transit-oriented development
- Make walking and biking easier
- Car share and carpool
- Improve fleet efficiency
- Achieve higher fuel-efficiency standards
- Switch to cleaner fuels
- Support intercity rail
- Improve freight movement

Reduced waste and industrial pollution
- Reduce, reuse and recycle
- Shift to alternative refrigerants
- Capture stormwater on-site

Adaptation
- Manage heat
- Pursue innovative cooling
- Protect air quality
- Manage stormwater
- Implement green urban design
- Preserve plants and trees
- Engage the public

Projected number of 100-degree days per year in Chicago

DERIVATION OF CARBON REDUCTION GOALS

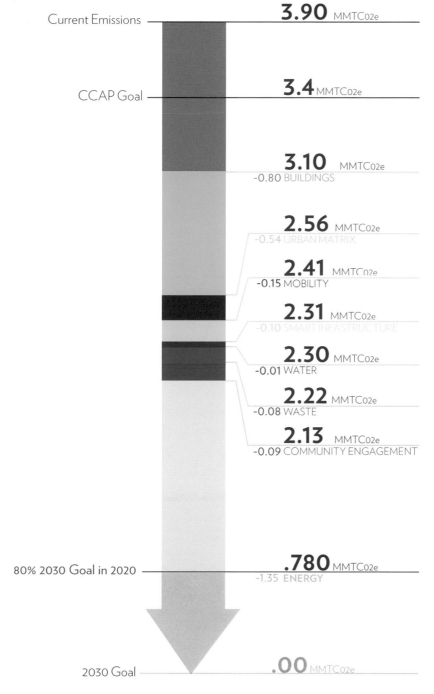

Current Emissions — **3.90** MMTC02e

CCAP Goal — **3.4** MMTC02e

3.10 MMTC02e
-0.80 BUILDINGS

2.56 MMTC02e
-0.54 URBAN MATRIX

2.41 MMTC02e
-0.15 MOBILITY

2.31 MMTC02e
-0.10 SMART INFRASTRUCTURE

2.30 MMTC02e
-0.01 WATER

2.22 MMTC02e
-0.08 WASTE

2.13 MMTC02e
-0.09 COMMUNITY ENGAGEMENT

80% 2030 Goal in 2020 — **.780** MMTC02e
-1.35 ENERGY

2030 Goal — **.00** MMTC02e

The Chicago Central Area DeCarbonization Plan goals are based on achieving the 2020 interim levels of reduction mandated by the 2030 Challenge. The 2030 Challenge is an initiative established by 2030 Architecture to achieve carbon-neutral building design for all new buildings and major renovations by the year 2030. The City of Chicago is a participant in the 2030 Challenge. Meeting the interim goal of the 2030 Challenge will also allow the city to fulfill its own goals set out in the Chicago Climate Action Plan. Since the study area for the DeCarbonization Plan is only a portion of the entire city, whenever reductions need to be made relative to the entire city, a prorated portion of the goal is determined by interpolation, using floor area as the ratio quantities.

Reductions by chapter

Each chapter of study uses a running total bar graph, similar to the one shown on the left, to indicate reductions from the strategies in the chapter, as well as the cumulative reductions of previous chapters. Since the 2030 Challenge goal can't be reached through the retrofit, conservation and waste reduction strategies outlined, an amount of renewable energy is proposed in the last chapter to make up for the cumulative goal shortfall. This method is not meant to suggest that renewable and distributed energy sources are not needed. Rather, it suggests a new way of thinking, and introduces a new method that does not allow renewable energy alone to be a panacea for the wasteful habits that are the cause of global warming, reversing the "just make more energy" patch that has been applied for the last century.

If we were to resolve the Loop's carbon emissions using renewable energy only, this is how much area and the systems it would require.

WIND FARM

Carbon to be made up	3.1 MMTCO2e
Equivalent kWh	4,447,632,712 kWh
Annual production for a 3 MW Vestas windmill	7,884 MWh
Number of windmills required	564 Vestas 3 MW windmills
Land area per windmill (10 rotor diameter spacing)	1,000,000 sm
Land area required	139,396 acres
Capital cost at $1.60/watt	$2.71 billion USD

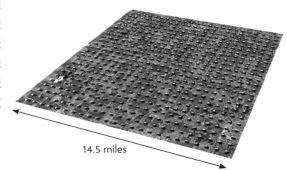

14.5 miles

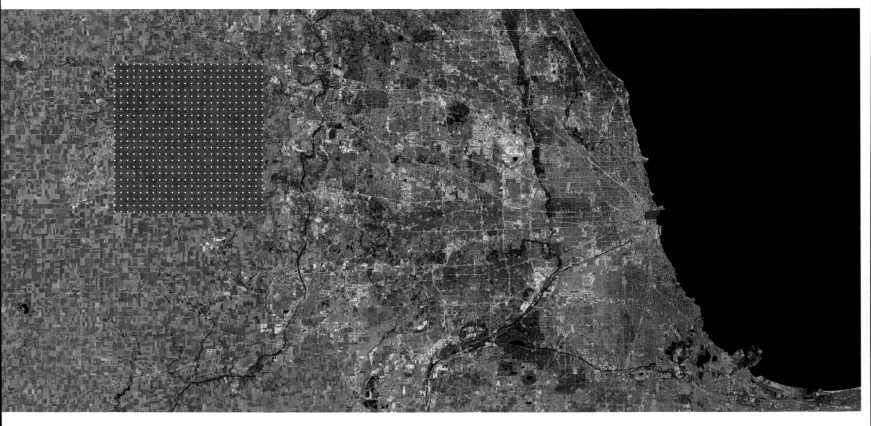

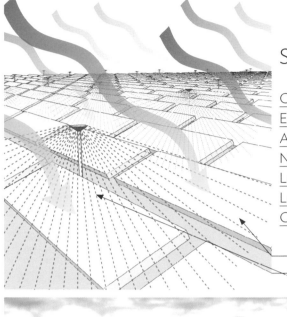

SOLAR ISLANDS

Carbon to be made up	3.1 MMTCO2e
Equivalent kWh	4,447,632,712 kWh
Annual production for a Sunpower 315 panel	0.3590 MWh
Number of panels required	12,388,949 Sunpower 315 panels
Land area per panel	0.001 acres
Land area required	12,389 acres
Capital cost at $6.80/watt	$27 billion USD

PV panels
Lake water sprinklers

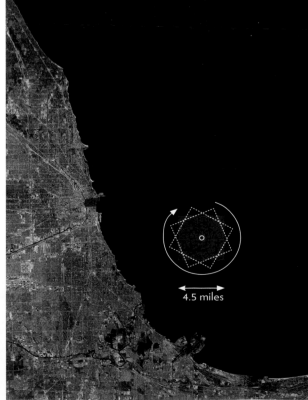

4.5 miles

BUILDINGS

In large cities such as Chicago, the energy demand on buildings can represent more than two-thirds of total carbon emissions. Modern lifestyle dictates the use of complex technological systems and conveniences in ways that significantly increase our reliance on power supply. But as we move further into the 21st century, there exists a strong opportunity for a paradigm shift: the transformation of buildings from power consumers to power generators.

BUILDINGS
Carbon Reduction Strategies

What if we could transform the existing building stock from a major part of the carbon problem into a contributing part of the carbon solution? In the United States, buildings account for an average of about 40% of carbon emissions. In Chicago, that number soars to 70%. This disparity is caused by the fact that our transportation footprint is relatively small. But as in many cities of similar age, Chicago's buildings often feature inefficient and outdated systems that cause high energy loads.

In this chapter, we first identify the **Carbon Reduction Goals** set by the Chicago Climate Action Plan for Buildings, and use these numbers as a basis to calculate goals for the Chicago Central Area DeCarbonization Plan.

In **Analysis**, we examine different building types that make up the majority of buildings in the Loop: Heritage (1880-1945), Mid-Century Modern (1945-1975), Post Energy Crisis (1975-2000) and Energy Conscious (2000-present), looking at the advantages and disadvantages for energy retrofits in each building type. Then we analyze building carbon emissions and energy use by examining the five building components responsible for the majority of energy use: building envelopes, lighting systems, HVAC systems, vertical transportation systems and plug loads. We then determine how modifications to those systems could improve building energy performance.

In the **Strategies** section, we look at why many buildings don't retrofit and examine strategies to encourage more buildings to do so, in order to reach overall carbon reduction goals.

We examine **Precedents** in Chicago and elsewhere in which buildings have gone through modernizations and conversions to become more energy efficient and to address diverse programmatic needs.

Finally, we identify a series of **Design Solutions and Pilot Projects** that offer a number of first steps toward addressing turning the existing Loop building stock into a more efficient neighborhood comprised of buildings that enjoy symbiotic relationships and produce energy.

CARBON REDUCTION GOALS

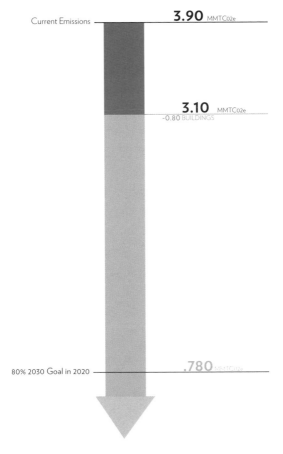

Current Emissions — **3.90** MMTC02e

3.10 MMTC02e
-0.80 BUILDINGS

.**780** MMTC02e
80% 2030 Goal in 2020 —

Specific goals for building emission reduction within the Chicago Central Area DeCarbonization Plan

New, energy-efficient buildings are opening every year, with many more under construction or in design. But new buildings account for only 1% of the entire building stock in the city, and remain only part of the solution. To address the millions of metric tons of carbon dioxide (CO_2) emitted by Chicago buildings every year, we must look at sustainable modernizations of Chicago's existing buildings.

The study area of the DeCarbonization Plan, the downtown Loop, accounts for many of the city's large and high-rise buildings. Though it covers less than 1% of the city's land mass, the Loop is responsible for 9% of its carbon emissions. This is partly due to building density, but building use is also a factor. Commercial buildings, which make up the majority of the Loop, typically have higher energy loads than residential buildings.

Overall, a combination of density, building size and building use contributes to the Loop's high percentage of building carbon emissions. The Chicago Climate

Action Plan (CCAP) stipulated that Chicago should reduce carbon emissions in half of the commercial building stock 30% by 2020, a total of the equivalent of 4.6 million metric tons of carbon dioxide (4.6 MMTCO2e) each year for the entire city.

The 2030 Challenge stipulates a much larger reduction of 80% in all new buildings and major renovations by the year 2020 and sets the goal of carbon neutrality by the year 2030.

For the Chicago Loop pilot area, it is proposed that the goal of a 30% average reduction be set for all buildings. To comply with this goal, the Loop needs to reduce building emissions by .8 MMTCO2e. Additional savings mandated by the 2030 Challenge would need to be made up using methods described in the following chapters (land use, transit, waste, water, etc.) as well as a significant amount of off-site renewable energy.

CCAP REDUCTION GOAL
4.6 MMTCO2e (30% in half of commercial buildings)

CLD REDUCTION GOAL
.8 MMTCO2e (30%)

LOOP AREA LAND MASS
1% of Chicago total

LOOP AREA CARBON EMISSIONS
9% of Chicago total

BUILDING MIX
90% Office
10% Mixed-use

EXAMPLE

At 4.5 million sf, Willis Tower comprises 1/20 of the Loop area, consumes 157 million kWh/yr, 1/1000 of total state energy, 1/20,000 of total US energy and 1/100,000 of the world energy at 1/10,000,000,000 of world land area.

RELATIVE CARBON FOOTPRINT OF BUILDINGS

Though the Loop comprises less than 1% of Chicago's total land area, it's responsible for 9% of the city's carbon emissions. When compared with three larger districts within the city (the Outer Central Business District, the Near Central Business District and the Urban Surround), the Loop's high carbon emissions per land mass are at least partly due to the amount of floor area per acre.

Central Loop

The City of Chicago is home to over 2.8 million people, and is the third largest city in the United States in terms of population. Chicago anchors a metropolitan area that is home to over 9.5 million people.

The total land area of the Chicago city limit is 234 square miles or nearly 150,000 acres.

Suburban Area

The edges of the city limit, especially to the west, are populated by densities similar to suburban areas. Buildings increase in typical size and density as one moves east and toward the central Loop area, where there are approximately 86 million square feet of buildings in only 480 acres.

9%

CARBON USE

1%

LAND AREA

CENTRAL LOOP
Land Area: 460 acres
Floor Area: 120,000,000 sf
Total Carbon: 3.9 MMTCO2e

Central Loop

CITY DATA MODEL

AS+GG developed a tool for visualizing building energy use on a citywide scale. The Data Model is intended to monitor and communicate all aspects of city buildings relating to their carbon impact. This serves a number of purposes: a method of analysis to inform energy-oriented planning decisions; an advertising platform for companies making efficiency upgrades to their buildings; an educational tool; and a forum to raise awareness about the carbon impact of the built environment and the efforts in place to mitigate or eliminate it.

As a planning tool, the power of the Data Model lies in its ability to synthesize large amounts of data into one holistic depiction of all the city's constituents and allow it to be viewed at once. In this study area, there are roughly 450 separate buildings. For each, there's a daunting amount of information that must be collected and organized. The Data Model condenses this matrix of figures into a simple, intuitive format. With this tool at his or her disposal, an urban planner can clearly discern patterns of energy use and efficiency to define problematic zones needing regulation or other intervention.

Marketing is a critical incentive for corporations adopting sustainable strategies. The Data Model is an interface designed to track consumption, but it can also emphasize improvement. This model will include a timeline function to track planned retrofits and estimate future performance improvements.

The timeline will serve to showcase the buildings that are making changes to their mechanical systems, facades, light fixtures and transportation policies that will create better operating efficiency. The timeline will highlight the resulting carbon reduction of these changes, for each building and for the city as a whole. This is a powerful way to attract attention and business to "green" corporations.

A broader purpose of this interface is simply the communication of these critical (though as yet unknown and uninvestigated) measurements of human impact. This medium reveals the hidden metabolism of our cities, making it impossible to ignore the incredible energy consumption of our structures. Public awareness is always the first step towards change, as well as the propelling force behind it. The Data Model hopes to be a catalyst.

INTERFACE

The web interface of the Data Model is its intended primary use. In the image shown here, the model displays buildings in various intensities of yellow, correlating to their total electricity consumption. Below the images in the interface are figures plotting the top 20 consumers, their gross floor areas, and the same data plotted against gross floor area. Buttons at the top of the page allow users to choose which data set is shown. To the right of these is the legend, defining the color scheme used for each data set.

 Carbon Emissions

 Energy Use Intensity

 Electricity Use

 Natural Gas Use

 Building Age

 Building Size

 Building Use

 Parking

BUILDING USE

Displaying this data set, the Data Model shows
the primary occupancy type of each building,
differentiated by color. A large percentage of the
buildings in this region have retail at street level, but
only the primary use is displayed for simplicity. The
color scheme is set by accepted international land
use color conventions.

Occupancy Type

- Office
- Government
- Mercantile
- Institutional
- Residential
- Hotel
- Mixed Use
- Parking

Building 123

123 W. Wacker Dr.

GSF=

1,300,000 ft²

FT²

- 1.1M

1.1M - 2.2M

2.2M - 3.3M

3.3M +

BUILDING SIZE

Gross square footage (GSF) of the buildings is represented here by color intensity. The greater a building's floor space, the darker its color appears in the model. We collected GSF data from sources including the Chicago Office of Emergency Management and Communication, real estate records and interpolation from the 3D geometry.

BUILDING AGE

The age of the buildings in the study area is represented here on a gradient scale from deep purple (oldest) through shades of pink to white. Note the lower height distribution of older (bluer) buildings. Building ages were compiled from GIS data provided by the Chicago Department of Buildings and from real estate records.

Year Built

 - 1900

 1900 - 1940

 1940 - 1975

1975 +

Building 123

123 W. Wacker Dr.

Year Built=

2011

**Building
123**

123 W. Wacker Dr.

Parking Spaces:

125

PARKING

Parking structures are marked here, with the size
of the icon indicating the relative capacity of each
location. Both stand-alone and integrated parking
are represented. Parking data was compiled from
interviews and online reservation sites.

31

ELECTRICITY USE

Total annual electricity use per building, in kilowatt hours, is visualized here by color intensity. The largest consumers are displayed in bright yellow. The values represented here are not corrected for building area square footage. Electricity use data was provided by Commonwealth Edison.

kWh / yr   (x 1000)

- 50,000

50,000 - 100,000

100,000 - 150,000

150,000 +

Building 123

123 W. Wacker Dr.

Electricity Use:

22,540,000

kWh / year

Building 123

123 W. Wacker Dr.

Natural Gas Use:

140,000

therms / year

Gas: Therms / year

- 250,000

250,000 - 500,000

500,000 - 750,000

750,000 +

NATURAL GAS USE

Total annual natural gas use per building, in therms, is visualized here by color intensity. The largest consumers are displayed in blue. The majority of buildings in the study area don't use natural gas, hence the predominance of gray. The values represented here are not corrected for building area square footage. Data was provided by Peoples Gas.

CARBON EMISSIONS

The carbon emissions number is intended to synthesize all of the energy expenditures of each building into one value. All of the factors that contribute to the operation of the building (electricity, natural gas, water, waste) are converted to a common metric, CO_2, using standard conversion rates as prescribed by the World Resources Institute. By this method it's possible to compare buildings by their total carbon impacts. This data is visualized using a red to green gradient. Green buildings have the lowest carbon to floor area ratio, so they're the the most efficient in their use of electricity, natural gas and water, waste production, and transit of their inhabitants based on average distances. Travel data is from the Center for Neighborhood Technology's Transportation Energy Intensity Index Study.

CO_2 / ft²

	- 9
	9 - 18
	18 - 26
	26 +

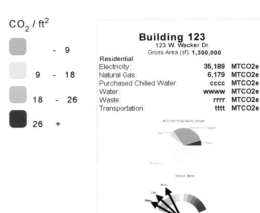

Building 123
123 W. Wacker Dr.
Gross Area (sf): **1,300,000**

Residential
Electricity:	35,189	MTCO2e
Natural Gas:	6,179	MTCO2e
Purchased Chilled Water:	cccc	MTCO2e
Water:	wwww	MTCO2e
Waste:	rrrr	MTCO2e
Transportation:	tttt	MTCO2e

©2009 ADRIAN SMITH • GORDON GILL ENERGY

(kWh + Therm) / ft²

- 9

9 - 18

18 - 26

26 +

ENERGY USE INTENSITY

While the carbon emissions metric combines all aspects of the building's use, the energy use intensity data set compares buildings only by their major energy expenditure—the direct use of electricity and natural gas in proportion to their gross floor area. This is a standard metric commonly used to assess building performance.

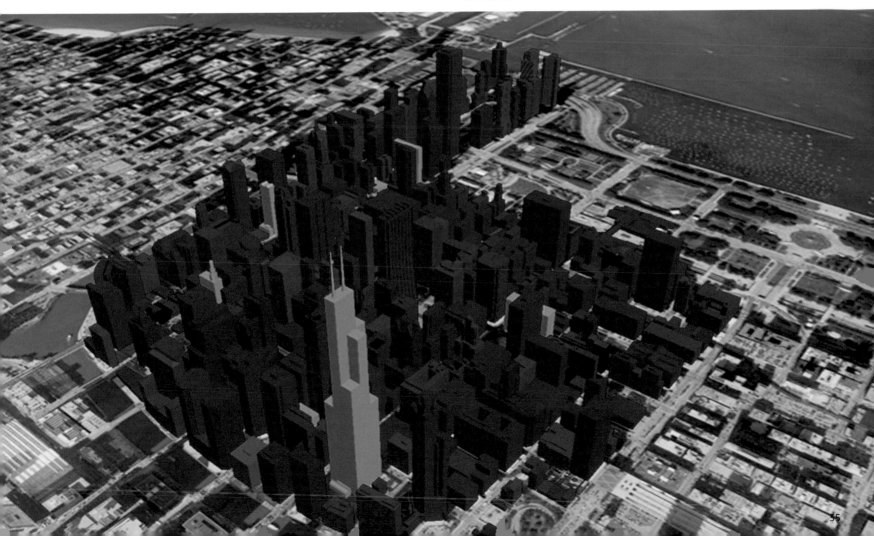

35

CHICAGO'S BUILDING TYPES

Chicago's skyline is among the most famous, and the most varied, in the United States. It's been said that the skyscraper was invented here in the decades after the Great Fire of 1871. Buildings in Chicago have often been categorized by architectural period, based on their materiality, size, layout and aesthetic expression. But if a set of energy reduction strategies is to be developed effectively, ultimately buildings need to be categorized by their "energy era" and their overall fitness in terms of performance.

HERITAGE
1880-1945

MID-CENTURY MODERN
1945-1975

POST ENERGY CRISIS
1975-2000

ENERGY CONSCIOUS
2000-present

Due to the historical development of technology in buildings, and how buildings have been used over time, there are many useful comparisons that can be made based on a building's energy era.

The first grouping is **Heritage Buildings**, which includes all structures constructed between roughly 1880 and 1945. This is a large span of years, but the key features these buildings have in common are their construction type (typically masonry, stone or terra cotta with punched windows) and their use, as originally designed, of natural light and ventilation.

The second grouping is **Mid-Century Modern Buildings**, which includes buildings constructed from about 1945 to 1975. After World War II, a revolution in architecture occurred, producing a new breed of high-rise buildings, densely lit structures of glass and steel. Curtain walls were developed to maximize the use of glass. More advanced HVAC systems allowed the building loads created by these large amounts of glass and dense lighting systems to be overcome. The disadvantage to these advances in architecture, however, was a sharp increase in the energy consumption of buildings.

The OPEC oil embargo of 1973 not only caused the development of stricter fuel-efficiency standards in automobiles; it also caused a new awareness of energy use in buildings, and greater investment in buildings' first cost to save energy over their lifespan. Large-scale buildings completed after 1975 often used technologies such as insulated glass, variable volume air systems and solar films or coatings that were developed to reduce energy loads and therefore operational costs. These **Post Energy Crisis Buildings** can be categorized as those built between approximately 1975 and 2000.

The final category of buildings are the **Energy Conscious Buildings** constructed during the current global movement toward energy efficiency, from approximately 2000 to the present day. The Chicago Energy Code was adopted in the year 2001; at the same time, green standards such as LEED became more common.

ENERGY REDUCTION IN BUILDING SYSTEMS

Financing Energy Retrofits
Many upgrades to existing buildings can be partially or fully financed through utility company incentives, energy service performance contracting, low-cost loans and tax-increment financing. Refer to the Funding chapter for more information.

The energy use estimates for each system outlined below will vary for each building based on its use, era and energy performance overall. By clearly categorizing the various systems that use energy, one can begin to understand opportunities for energy savings.

In this model of the Central Loop, the various energy eras of buildings are illustrated. Buildings are examined by the portion of energy used by each building system. The analysis illustrates the energy use in a poorly performing building versus today's standards and energy codes, and what's possible for a new high-performance building. The energy use of an older system can be three to five times more than that of a new, high-performance system.

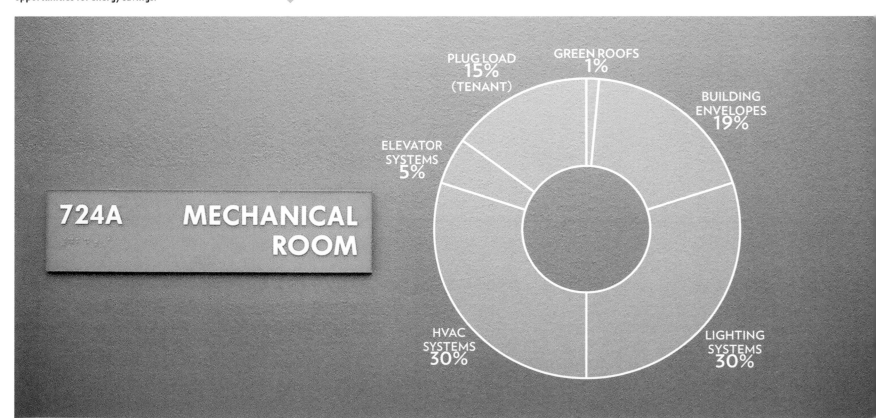

PLUG LOAD
15%
(TENANT)

GREEN ROOFS
1%

BUILDING
ENVELOPES
19%

ELEVATOR
SYSTEMS
5%

724A   MECHANICAL ROOM

HVAC
SYSTEMS
30%

LIGHTING
SYSTEMS
30%

HERITAGE   1880-1945

Heritage buildings feature strong opportunities for natural ventilation and daylight because of shallow lease spans, as well as strong thermal mass for heat absorption. However, many of these buildings haven't benefited optimally from changes to building systems over the years. In some cases, landmark status can also prevent these buildings from receiving the benefits of full sustainable upgrades.

Old Colony Building. Architect: Holabird & Roche

Typical Exterior Wall
- Brick, stone, terra cotta (uninsulated)
- Double-hung windows with single-pane, clear glass
- Glass percentage: 25 to 50%

Typical Mechanical Systems
- HVAC has generally been retrofitted or completely replaced over the years, especially for buildings older than 1930. Therefore a large variety of mechanical systems, from radiators to fan coil units to variable-volume overhead systems, exist in these buildings. Often systems don't operate optimally because of changes over the years. Cooling systems have likely been added to the original building, as air conditioning was not widely used prior to 1950.
- Many heritage buildings have opportunities for natural ventilation from short lease spans or light wells, but sometimes these have been covered over.
- Some buildings use district heating and cooling due to lack of plant capacity. Others have large rooftop units added.

Typical Lighting and Electrical
- Opportunities exist for natural light through short lease spans and light wells, especially in the oldest buildings.
- Buildings that did not originally have electric lighting have been retrofitted. However, early lighting retrofits typically do not have as many lights as the overlit buildings from the 1950s through the 1970s. Therefore, lighting and electrical loads in older buildings often consume less energy than in later buildings.

Typical Layout
- Shallow lease spans and narrow floor plates, or large floor plates with one or more light wells.

Site and Environment
- The lakefront was protected by the early planners of Chicago. However, architects of the first high-rise buildings and early zoning regulations did not yet have a strong awareness of site improvements with positive environmental impacts, such as setbacks and green or permeable areas.

MID-CENTURY MODERN

1945-1975

Mid-century Modern buildings typically feature large amounts of glass due to the development of curtain walls. In this era, technology changed the way buildings were made. The use of the natural environment for heating, cooling and lighting was largely abandoned in favor of artificial control of interiors. Due to trends in the use of office space, dense lighting and large HVAC systems became common. Mass production speeded up construction, reduced cost and increased modularity in buildings.

The Federal Center. Architect: Mies van der Rohe

Typical Exterior Wall

- Curtain wall systems: steel, aluminum, stone, concrete
- Windows are commonly clear or slightly tinted (green, gray or bronze) glass with manual shading devices. Most curtain walls were built before the development of thermal breaks, and most glass is single pane, as insulating units were only in their early development.
- Glass percentage: 50 to 80%

Typical Mechanical Systems

- Mechanical systems are typically induction or fan coil units at the perimeter with interior air via ducted systems. Constant electric reheat is the typical interior system; in many buildings, fortunately, it has been phased out in favor of a more efficient variable volume system.
- Mechanical plants were also originally constant volume, but many have been retrofitted for variable volume, and the use of economizer cycles.

Typical Lighting and Electrical

- Lighting density in the mid-century was often as high as 5 watts per square foot, and buildings were designed as "heat by light." Over the years, with the development of computers, it was found that these high levels were not necessary, and it was more efficient to use the HVAC system for heating. Therefore later tenant fit-outs often reduced the lighting density.
- Equipment loads (or "plug loads") began to increase as new technology was introduced.

Typical Layout

- Large floor plates of 30,000 to 50,000 sf with a central core.

Site and Environment

- Large public spaces, allowing more sunlight in the city, were common in the mid-century era, but they were often made of hard surfaces such as granite paving, which negatively affects stormwater management and the urban heat island effect.

POST ENERGY CRISIS   1975-2000

The energy crisis of the 1970s affected building construction techniques. In high-rise buildings, insulating glass became common, solar coatings were developed to reduce heat gain, HVAC systems became more efficient and the amount of lighting was reduced. With the development of computers, however, internal plug loads in buildings increased.

The Thompson Center. Architect: Murphy/Jahn

Typical Exterior Wall

- Curtain wall systems: aluminum, stone, concrete
- Up to 1990s: mirrored or dark-tinted insulating glass
- 1990s-2000: clearer glass with Low-E coatings
- Glass percentage: 60 to 80%
- Introduction of thermal breaks

Typical Mechanical Systems

- Induction or fan coil units at perimeter with interior air via ducted variable volume systems.
- Loads decreased as less heating was needed along the perimeter glass walls. Mechanical plants begin to use variable equipment.

Typical Lighting and Electrical

- Lighting steadily reduced from 1970s levels as heat-by-light incentives were phased out. Partial or complete retrofits/de-lamping may have decreased lighting loads.
- With the development of computers, office space became much more energy intense in terms of plug loads. This often caused overall electrical energy use, especially in office buildings and trading floors, to increase sharply.

Typical Layout

- Large floor plates of 30,000 to 50,000 sf with a central core.

Site and Environment

- With increased zoning requirements for landscape, buildings after 1975 tend to be surrounded by more trees and have a site design that is sensitive to pedestrians. New city ordinances raised awareness of stormwater management.

Renewable Energy

- After the energy crisis, ideas about integrating solar energy generation into buildings were being developed. But building-integrated renewable energy was not generally used in the Central Loop due to long payback periods.

ENERGY CONSCIOUS   2000-PRESENT

Energy codes fundamentally changed the way buildings are designed. Today's architects and engineers must have a constant awareness of building behavior as it relates to environmental impact and energy requirements. Owners and tenants are beginning to demand high-performance buildings. Buildings are slowly regaining their relationship to the external environment, which is now balanced with technology that optimizes human comfort.

111 South Wacker. Architect: Lohan Caprile Goettsch

Typical Exterior Wall
- Curtain wall systems: aluminum, stone, concrete
- Insulated glass with Low-E coatings is the industry standard in the United States.
- As energy codes become stricter, glass percentages become lower.
- Triple glazing and double-skinned walls are introduced.

Typical Mechanical/Electric/Plumbing Systems
- VAV or fan coil systems with less perimeter heat needed due to better envelope performance.
- Use of economizer cycles, variable frequency pumps, drives and digital controls.

Typical Lighting and Electrical
- Lighting levels reduced to meet energy codes and LEED. Natural daylight and light sensors are the industry standard to meet energy guidelines. New technologies such as compact fluorescent lighting and LEDs are becoming more common.
- Plug loads remain high, but more efficient equipment is being developed to save energy.

Typical Layout
- Large floorplates remain common for office space, but more attention is paid to orientation and shallower lease spans for natural light.

Site and Environment
- New environmental awareness has encouraged the use of permeable surfaces, urban parks and stormwater retention. When selecting a site, the rehabilitation of brownfield areas and pollution cleanup are encouraged. High development density and mixed-use buildings in urban areas allow better access to cleaner public and bike transportation, as opposed to cars. The importance of public parks remains high, enhanced by the development of Millennium Park in recent years.

Renewable Energy
- Buildings are beginning to integrate energy generation elements such as solar panels and wind turbines. Tax incentives are making these elements more financially viable.

BUILDING ENVELOPES

The performance of the building envelope typically affects 15-30% of overall energy use. A high-performance building envelope relies on a strong thermal resistance. Many new window systems can perform at extremely high thermal resistance levels, making them incredibly efficient. Typically, however, cost increases with higher-performing exterior envelopes, which prolongs the product's payback period. Building envelopes are an important consideration in modernization and sustainable upgrades.

Funding

Full envelope upgrades often require long-term loans. However, smaller elements such as window film, seal improvement and gasket improvement can be financed as part of short-term capital expenditures or short-payback projects through performance contracting. Refer to the Funding chapter for more information.

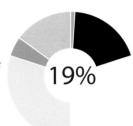

19%

R-Values relative to system options:

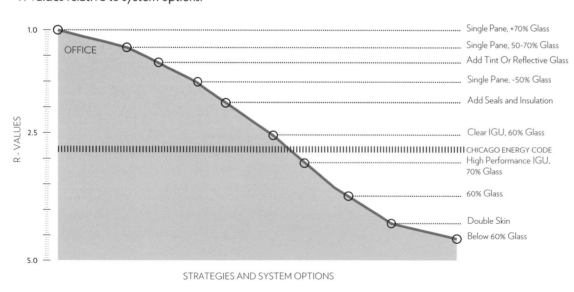

OFFICE

R - VALUES

STRATEGIES AND SYSTEM OPTIONS

- Single Pane, +70% Glass
- Single Pane, 50-70% Glass
- Add Tint Or Reflective Glass
- Single Pane, -50% Glass
- Add Seals and Insulation
- Clear IGU, 60% Glass
- CHICAGO ENERGY CODE
- High Performance IGU, 70% Glass
- 60% Glass
- Double Skin
- Below 60% Glass

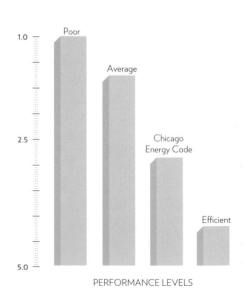

PERFORMANCE LEVELS

Poor

Average

Chicago Energy Code

Efficient

Thermographic image of a masonry building in the Loop showing heat loss through glazing ▶

LIGHTING SYSTEMS

Lighting systems have advanced greatly from the original gas fixtures used in Chicago's early buildings. Today's lighting systems can be designed and automatically controlled based on the optimal amount of daylight for the human eye. However, many mid-century buildings are over-lit. In that era, the necessary lighting density was thought to be as much as five times what's common today. Lights were also used to help heat the building.

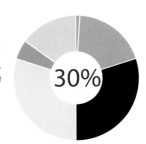

Funding

Lighting upgrades are typically a short payback item and can be eligible for incentives through local utility companies. For more information, refer to the Funding chapter.

Wattage relative to system options:

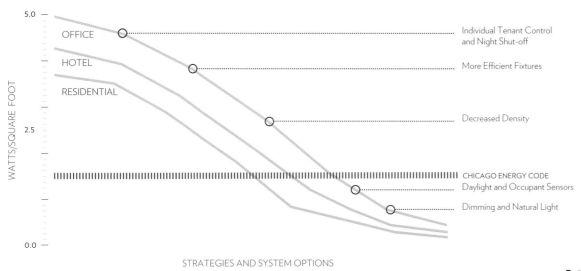

OFFICE

HOTEL

RESIDENTIAL

WATTS/SQUARE FOOT

Individual Tenant Control and Night Shut-off

More Efficient Fixtures

Decreased Density

CHICAGO ENERGY CODE
Daylight and Occupant Sensors

Dimming and Natural Light

STRATEGIES AND SYSTEM OPTIONS

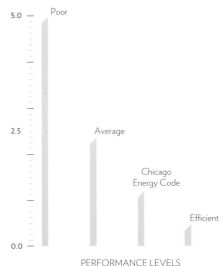

Poor

Average

Chicago Energy Code

Efficient

PERFORMANCE LEVELS

Buildings with deeper lease spans and less daylight require more electric lighting. Lighting energy is also affected by the use of the building and whether the lights are automatically controlled. Lighting can represent 15-35% of the overall load, depending on its efficiency and how much daylight is being used in the building.

HVAC SYSTEMS

Mechanical system sizes are driven by the internal loads on the building (heat from people, lights and equipment) as well as the external envelope loads. Therefore, the mechanical system efficiency is ultimately dependent on the architectural design of the building and its use. However, mechanical systems have become more efficient over the years in the ways they're able to react and adjust to different conditions through efficient distribution of air, variable volume operation, digital control to sense temperature changes and use of outside air when weather permits.

Funding

Mechanical upgrades have various financing methods based on how extensive the work. Minor improvements such as variable frequency drives can be partially covered by utility incentives, while longer payback renovations may rely on long-term loans or tax increment financing. Refer to the Funding chapter for more information.

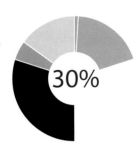

Kilowatts per hour relative to system options:

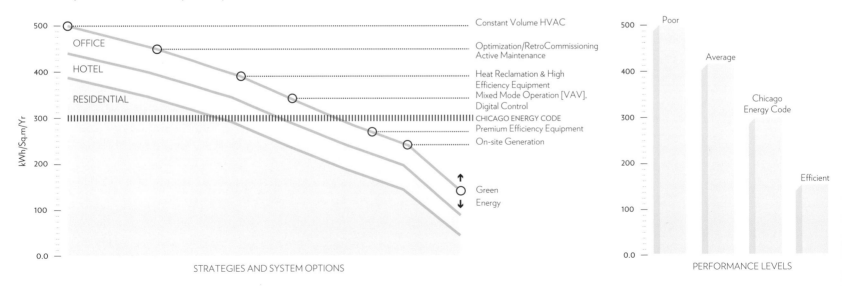

Mechanical equipment (excluding envelope loads) can account for about 15-30% of overall energy use. This includes the energy needed to distribute heating and cooling throughout the building.

VERTICAL TRANSPORTATION SYSTEMS

Energy used by elevators and escalators can represent between 2 and 10% of the energy in a typical Loop building. This depends on building size, number of elevators and building use (including frequency and amount of elevator traffic). Over the years, typical vertical transportation equipment has become about 50% more efficient, but many old systems are still in operation. New technologies have also been developed to make elevator systems smarter by using digital controls to reduce the number of trips.

Funding

Elevator upgrades are most often paid for as a planned capital expenditure when they reach the end of their life. The incremental cost of high performance systems can also be assisted by utility incentives. Refer to the Funding chapter for more information.

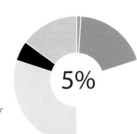

5%

Kilowatts per hour relative to system options:

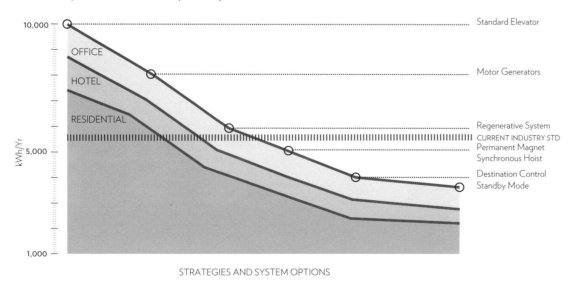

STRATEGIES AND SYSTEM OPTIONS

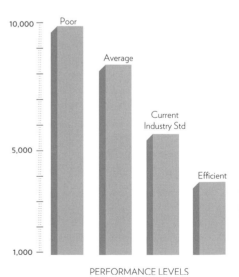

PERFORMANCE LEVELS

Vertical transportation energy represents between 2 and 10% of overall energy in a typical Loop building. Office and institutional buildings tend to have the greatest energy use from elevators due to their dense occupancy. Supertall buildings such as Willis Tower have more than 100 elevators.

EQUIPMENT OR PLUG LOAD

Equipment loads in buildings can most simply be understood as the electricity needed for anything that needs to be plugged in to an electrical outlet. While most loads in buildings (lighting, envelope, etc.) have decreased over time, plug loads have increased, especially since the 1980s when computers became widely used. Measures that can be taken today to reduce equipment loads include choosing Energy Star (or better) equipment and automatic shutdown. Future technology will be able to reduce these loads further as electronic equipment becomes more advanced and consumes less energy.

Funding

Building policy is integral in achieving a phased reduction of plug loads since these loads are reliant on tenant behavior. Building owners can help by developing building standards, green publicity programs and and incentives for tenants. Refer to the Funding chapter for more information.

Wattage relative to system options:

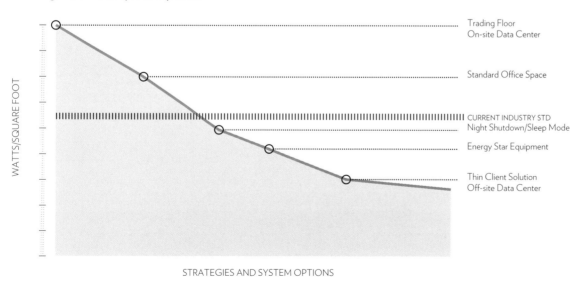

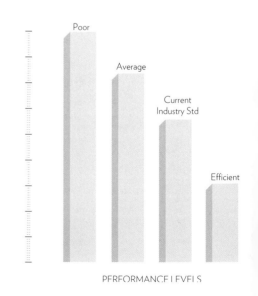

Equipment loads vary widely based on building use. For example, a residential building may have only 15% of its loads from equipment while a trading floor, industrial or broadcasting building may have over 50%.

GREEN ROOFS AND ROOF INSULATION

Loop buildings tend to have a relatively low proportion of roof area, but roof insulation and the use of low-albedo or green roofs are still important for reduction of the urban heat island effect, which tends to increase the localized temperature of the Loop. Older, dark-colored roofs contribute to this effect, as they can reach nearly 200 degrees Fahrenheit in the summer. Green roofs are also important in holding stormwater, which reduces carbon emissions by reducing the need for water treatment.

Funding

All projects in Chicago that apply for planned development zoning or tax increment financing are required to have green roofs. Refer to the Funding chapter for more details.

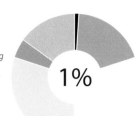

1%

Roof insulation relative to system options:

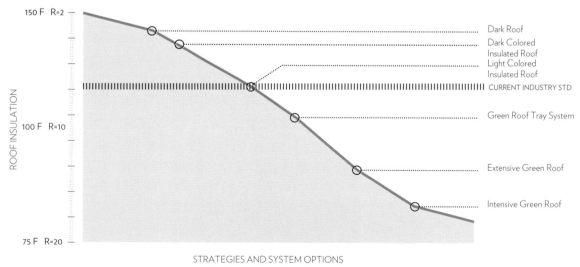

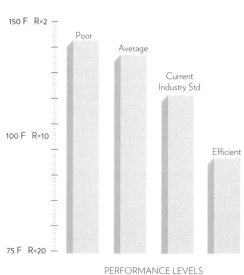

Roof area represents a small portion of envelope loads for most Loop buildings, but the use of green roofs is valuable in terms of potential benefits such as reducing stormwater runoff.

ENERGY SOURCES

Carbon emissions in buildings can't be examined without considering the source of a building's energy. A significant percentage of the energy from power plants is lost as it travels over the power grid. Peaks in energy loads not only affect cost, but also tend to increase the energy sourced from coal-fired power plants. Controlling and storing energy in buildings to reduce these peaks will allow buildings to contribute less to carbon emissions. On-site energy generation can reduce transmission losses, while renewable on-site generation can allow buildings to operate carbon-free.

Funding

Financing assistance for alternative energy sources is available at the state and national levels. Refer to the Funding chapter for more information.

Carbon Intensity relative to system options:

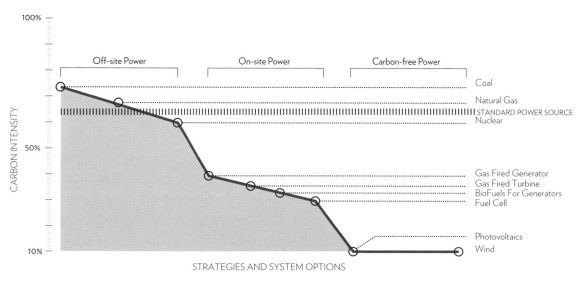

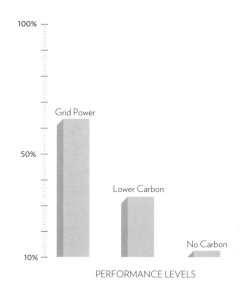

All energy in a building comes from a source, typically the power grid. In the Loop, most power is sourced from nuclear plants and natural gas. But there is still a strong reliance on coal-fired electric power in the region, as well as a need to reduce peak loads.

WHY EXISTING BUILDINGS DON'T RETROFIT

Lack of consistent benchmarking and comparison system

Currently buildings rely on voluntary systems such as Energy Star to benchmark against other buildings. Energy Star is best used for office buildings, but is still developing as a tool for unique occupancy uses. It does not address how existing buildings compare to current codes or other buildings. Therefore, there is little competition or publicity regarding energy savings or improvement.

Limited energy code requirement for existing buildings

While energy codes and ASHRAE standards are growing increasingly stringent for new buildings, there is currently no minimum performance level for existing buildings. Only major renovations are required to comply, and those renovations only to a limited extent. The financial burden of requiring older buildings to comply with the new code is likely too great, but there is potential for phased implementation of a less stringent energy standard, and use of incentives as buildings renovate, to help improve the performance of existing buildings. There is also an increasing trend for projects with public funding to have required energy standards.

Payback periods

Short payback projects (generally less than 10 years) are financeable through energy service companies (ESCOs) and through ComEd/DCEO incentives (though some of these incentives require even shorter paybacks of three to seven years). Longer payback projects are often not performed because of financing difficulties.

Inability to sub-meter

Many commercial buildings are not metered for individual tenant lighting, and installing such metering has a high capital cost. Also this often requires a change from 277 to 480 volt power.

Lack of incentive due to cost pass-through to tenants

Many large commercial and residential buildings pass along base building costs to tenants through leases or assessments. Therefore the incentive to improve energy efficiency is absent. However, building tenants are becoming increasingly aware of these costs, and a more competitive market may offer future advantage to buildings with lower pass-through utility rates.

Lack of personal and professional commitment and coordination

Some buildings do not have the advantage of a proactive or experienced management and operational staff. Other times, operational staff have energy-saving recommendations, but lack the coordination with financial initiatives necessary to obtain funds.

Lack of corporate support and business community engagement

A lack of corporate support can stall energy-saving projects and investment.

Financial challenges

Some building owners simply do not have the financing available or the ability to get the loans necessary to pay for large-scale energy projects. Energy projects with large amounts of savings are often too invasive or large-scale to be paid for as a planned capital expenditure, and are better done as part of a larger renovation project. But getting these types of projects financed can be difficult if credit is not available or owners are not able to raise the capital required to repay large loans on an annual basis.

Many late 1970s and early 1980s office buildings have only reduced a small amount of their energy use since construction. Available funds are often prioritized towards more visible areas such as lobby renovations.

The IBM Building is undergoing a change of use from office to hotel. These types of renovation can lend themselves to significant energy upgrades at a low additional incremental cost. However, older buildings are not required to upgrade to today's energy codes during such renovations.

POTENTIAL STRATEGIES TO ENCOURAGE RETROFIT

The landmark Inland Steel building has done extensive study of potential retrofits including envelope upgrades and improvements to tenant spaces. Only partial implementation has taken place to date due to the challenging scale of the proposed project.

The Chase Tower has upgraded mechanical and lighting systems as part of an energy retrofit project with a short payback period. The single-glazed exterior wall has not been upgraded due to a long payback period.

Consistent benchmarking and comparison system

Create a policy whereby existing buildings must benchmark their energy use intensity. This will identify the buildings with the most energy savings opportunity and force managers and users of low-performing buildings to realize and report how much energy they are using. Example: The Green Office Challenge (though it's voluntary).

Create an energy code requirement for existing buildings

Create a minimum performance level for existing buildings. This could be targeted to the largest buildings. Due to the cost burden for existing buildings to upgrade, this requirement should be accompanied by an incentive system and/or a gradual phasing. Example: NY initiative, California codes, European codes.

Financing for longer payback periods

Encourage policies that allow for loans and incentives to apply to projects with a payback up to 25 years. Educate building owners about incremental cost and maintenance cost, which can be considered in order to reduce payback periods.

Implement separate metering for all tenants

Provide policy that encourages base buildings to meter separately. Offer incentives to solve cost issues related to the installation of the meters.

Allow tenants to pay utilities directly, and also encourage better tracking

Provide policy to encourage large buildings to meter separately in order to ensure that tenants pay actual lighting and heating bills. Encourage green leasing practices where tenants will become more aware of their usage, and may require a minimum performance level for the base building.

Increase personal and professional commitment and coordination

Train operational and management staff in green practices. Hold regular meetings for communication between those responsible for financing and those responsible for operations. Be proactive about energy audits. In large buildings, appoint a staff member responsible for energy savings.

Increase corporate support and business community engagement

Increase support both within large organizations and across organizations through programs such as the Green Office Challenge. Inspire competition between corporations to make saving energy and reducing waste more attractive.

Overcoming financial challenges

In the Funding chapter, we take stock of the reasons why building owners don't retrofit and offer a number of suggested financing strategies for existing building owners to use. It's suggested that pilot projects apply for competitive grants from the U.S. Department of Energy so that the benefits of these pilot projects can be replicated throughout the Loop study area. A comprehensive approach is recommended to marshall grants (for the early stages of capital needs, such as energy audits, commissioning, education and outreach efforts) and later-stage capital sources, for equity to support project financing. This financing will use a variety of tools including tax increment financing, special service areas, loans and existing Commonwealth Edison programs.

WHAT'S BEEN DONE

Bank of America building

Environmental Systems Design, Inc. (ESD) worked on a retrofit of the 1.2 million sf, 44-story Bank of America building (1934) in Chicago. Improvements included updates to standby and emergency life safety generation systems. The installation included required generator and related life safety systems to comply with the 2000 Chicago Electric Code. ESD also provided lower tower electrical distribution engineering upgrades, emergency generator design, and other HVAC, plumbing, energy and elevator studies and improvements.

ESD provided LEED commissioning services and detailed retro-commissioning and re-engineering. The commissioning process produced annual energy savings of more than $300,000 in low-cost immediate energy conservation measures. The process also identified long-term energy conservation measures to be implemented as space renovations occur, anticipated to result in additional energy savings of $900,000 per year.

Chase Tower

ESD was also commissioned to develop, design and implement a facility master plan for the renovation and upgrade of the mechanical, electrical, plumbing, life safety, security and controls system infrastructure of the 2.4 million sf Chase Tower. The primary goal was to position the renovated building as "best-in-class" for another 50 years by improving and developing new contemporary public and tenant amenities while significantly reducing overall energy consumption.

Significant renovations included the perimeter HVAC system, toilet room plumbing systems, a chiller plant retrofit, VAV conversion of 40 existing air handling systems, 12kV electrical riser redistribution, tenant electrical lighting and power improvements and a retrofit of the entire building automation system. Using advanced modeling techniques including DOE II and EnergyPlus energy simulations, ESD modeled the post-retrofit energy savings to be over $1 per square foot (over $2.4 million per year). ESD provided the design of a robust building infrastructure communications riser (BIR), building automation system (BAS) and heating and lighting control systems. The BIR required primary and redundant fiber-optic cabling between high-quality IT industry smart switches. Cat5E cabling from the smart switches was installed to each IP addressable field controller, server and operator workstation. The BIR provides communications for the BAS, heating and lighting control systems. The upgrade cost about $1.3 million.

The replacement BAS design required all local controllers to utilize BACnet or LonMark standard communication protocols, Ethernet communications from field controllers to a central file server and a standard web browser over the Internet to provide remote operator access. The replacement BAS has 3,500 physical control points and cost about $2 million.

The Bank of America building in Chicago underwent a sustainable modernization that resulted in significant energy conservation.

Chase Tower's mechanical and lighting systems renovation allowed the building to save $900,000 per year in energy costs.

WILLIS TOWER

The Willis Tower (formerly Sears Tower) has saved more than 30% of its energy since construction in 1974 by using variable volume air, more efficient controls on equipment and higher-efficiency light fixtures.

Completed strategies

- Re-lamping to T8, courtroom lighting renovation
- Change to VAV boxes from CAV
- Various operational practices
- Replacement of leaking plumbing fixtures
- Mechanical plant operational tuning
- Variable speed drives on pumps and fans

Energy savings

- 50% of electricity
- 32 million kWh reduction, 1991 to 2008
- 22 watts/sf current energy intensity (not including gas heating)
- 61% reduction in water use, 1990 to 2008

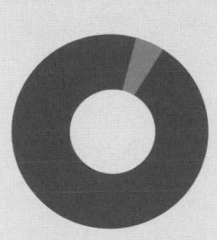

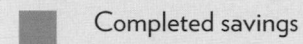
CCAP Loop goal (after incentives)
540 million kWh

Completed savings
- 6% of CCAP investment goal
- Building area
- 69% of Loop sf

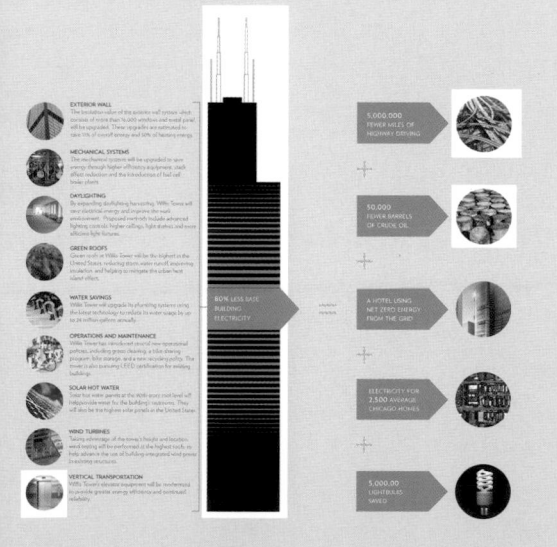

Willis Tower is undertaking a new greening and modernization program with the goal of saving up to 68 million kWh annually, including upgrades to mechanical systems, plumbing, elevators, lighting and the exterior wall.

Strategies: Energy upgrade (1990-2008)

- Re-lamping, VAV upgrade, operational upgrade

Strategies: Modernization (planned 2009-2013)

- Lighting control upgrade
- Re-glazing with double or triple pane
- Envelope insulation upgrades
- Mechanical plant modernization
- Elevator modernization
- Stack effect mitigation
- Water efficiency upgrade
- Green roofs
- Renewable energy
- Induction unit removal

Energy savings

- 30% savings to date, 45% more savings planned
- Total savings: more than 130 million kWh
- 35 watts/sf current energy intensity—19 watt/sf goal

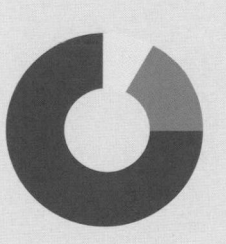

Planned savings
- 9.7% of CCAP investment goal

Completed savings
- 15% of CCAP investment goal

CCAP Loop goal (after incentives)
540 million kWh

DESIGN SOLUTIONS AND PILOT PROJECTS

It's important to encourage all buildings to realize savings using incentives already in place. However, the energy reductions needed to reach the goals set out by the Chicago Climate Action Plan and the 2030 Challenge are too aggressive to simply rely on existing incentives and on building owners to take the initiative to retrofit. More leadership and more incentives will be required to reach these goals. Therefore, the next necessary step is to begin a trend of retrofitting buildings in the central Loop, based on the 30% savings target, by developing groups of buildings to work together as pilot projects.

- To have the greatest chance of making a large impact in the target area, a key group of 83 buildings has been identified to create a pilot project, based on the largest square footages, owners and operators. If this group of buildings (about 70% of Loop square footage) can attain a 30 to 40% reduction, this would exceed the targets set forth by the CCAP. The Loop should be a leader in reaching the CCAP goals; therefore the reduction goal of .8 million metric tons of CO2 for the target area set forth in this chapter is more aggressive than the average level of the citywide CCAP goal, which assumes a 30% energy savings in half of all buildings.

- Options for pilot projects include the grouping of buildings into energy districts based on their commonalities in use, location or era. Dividing buildings into target districts would be a sensible way to allow for smaller, more focused projects.

- The savings potential of these pilot buildings is currently being studied and categorized according to their existing performance and the reduction methods described earlier in this chapter. As each building is examined in more detail, the pilot groups can be further refined and coupled with financing methods that are unique to that group of buildings, as discussed in the Funding chapter.

- Retrofitting buildings is not only critical for the overall carbon emissions of the city; it also renovates and upgrades the building stock so that many older buildings can remain financially viable for decades to come. Once older buildings upgrade to new green standards, they will be able to attract the new tenants and residents that are now demanding more healthy, comfortable and efficient space.

ENCOURAGE SELECT MAJOR RENOVATIONS
30-40% savings in pre-1975 buildings

Although not all buildings are in a position to undergo major renovations, a surprisingly large number will be ready in the coming decade. This is because many mid-century buildings have systems that will soon reach the end of their useful life and require replacement. There are also many office buildings undergoing changes in tenancy that will require office fit-outs or conversions to other uses, which will enable extensive renovations. This large number of buildings ready for renovation provides the opportunity for a coordinated, systematic response to the renewal of these systems in which energy efficiency is the overarching goal.

If these major renovations are energy-conscious, a 35% average carbon reduction goal can be realized for many large buildings.

Based on our data model, 90% of the Loop square footage was designed and constructed before the 1975 energy crisis. These buildings have great opportunity to partner with the city on their planned renovation programs to reach the aggressive energy target.

Solutions
· Lighting and lighting control upgrades
· Mechanical system retrofit or replacement
· Building envelope upgrades or glass replacement
· Elevator modernizations
· Commissioning, tuning or re-engineering equipment
· Stack effect mitigation

Existing initiatives
· TIF Program

Potential impact
· The largest 10 buildings (or complexes) in the Loop account for about 25 million sf or 14% of the square footage. If each one of those buildings saved 30% of its energy, nearly half of the Loop CCAP goal would be attained. If all of the 83 pilot buildings renovated to a 30% savings level, .6MM tons of CO2 would be saved, which is more than one-third of the entire city's CCAP goal for buildings.

RETROFIT THROUGH LOW-COST MEASURES
15-25% savings in post-1975 buildings

A 15% average savings for the remaining Loop buildings, in addition to the major renovations described on the previous page, would attain the overall carbon reduction goal for buildings of **.8 million metric tons of CO2**. This total would mean that the Loop alone could achieve about half of the overall CCAP citywide savings goal for buildings. The advantage of using low-cost, lower-savings measures is that virtually all buildings can participate. However, undertaking these programs requires dedication on the part of building management and owners, and on the availability of loans or allocation of existing capital budgets to energy upgrades, since the existing ComEd/DCEO incentives do not cover all expenses. Reduction requirements along with increased incentives would be necessary using this method, because all buildings would need to participate. Precedents, such as New York City's new requirement that all buildings over 50,000 sf have some level of energy renovation, could inform a similar program in Chicago. The current system does not require existing buildings to comply with energy codes put in place after they were built. These buildings can therefore continue to consume as much as four or five times more than a code-compliant new building, without any requirement for even the most non-invasive, short-payback energy upgrades.

Solutions
- Encourage or require buildings to track usage and energy intensity, making the data public for peer buildings and the city.
- Expand ComEd/DCEO incentives for, or require, simple prescriptive upgrades such as T8 lamps and variable volume pumps, regardless of whether a building is undergoing a renovation project.
- Incentivize or require a building diagnostic study and commissioning of mechanical plants.
- Incentivize or require tenant and building occupant programs that encourage behavioral changes.
- Incentivize or require programs whereby large building owners perform maintenance programs such as exterior wall seals, filter changes, equipment tuning and cleaning.
- Encourage buildings to pursue energy performance contracting for minor, short-payback upgrades.
- An effective combination of requirements and incentives is critical for total participation. In some cases, the requirements may be a feature of the incentives.

Existing initiatives
- Green Office Challenge
- Green Hotels
- Chicago Green Restaurant Co-op
- Energy Service Contracting/Clinton Climate Initiative

Potential impact
- If all buildings in Chicago were able to save an average of 15% of energy by 2020, the carbon reduction goals of the CCAP could be attained. This level of savings should be possible at low cost and/or short payback for most buildings (without major renovations).

EXAMPLE ENERGY DISTRICT BY BUILDING SCALE

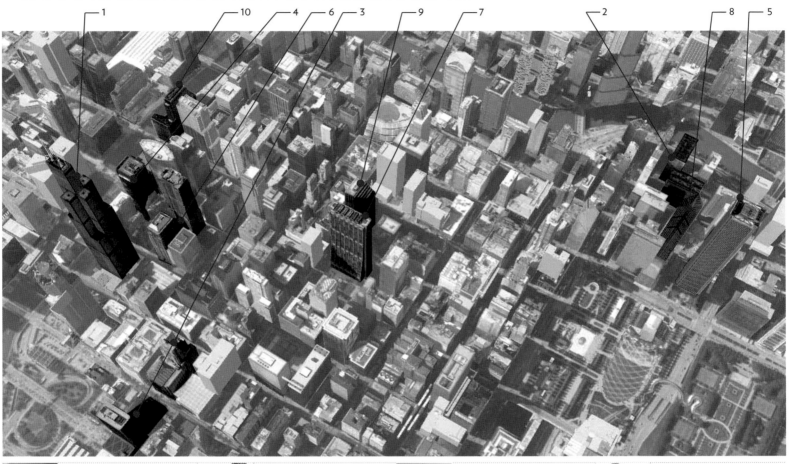

Building Name : **Willis Tower**
Address : 233 S. Wacker
Year Built : 1974
Architect : SOM
# of Stories : 108 Stories
Occupancy : Office
Area : 4,300,000 GSF

①

Building Name : **Illinois Center**
Address : 111 E. Wacker
Year Built : 1970-1972
Architect : Mies van der Rohe/Fujikawa
# of Stories : 2x30 Stories
Occupancy : Office
Area : 3,700,000 GSF

②

Building Name : **Chicago Board of Trade**
Address : 400 S. LaSalle
Year Built : 1945/1983
Architect : Holabird & Root/
# of Stories : 44 Stories
Occupancy : Office
Area : 3,200,000 GSF

③

Building Name : **Chicago Mercantile Exchng.**
Address : 10-30 S. Wacker
Year Built : 1983-1985
Architect : Fujikawa, Johnson & Assoc.
# of Stories : 2x40 Stories
Occupancy : Office
Area : 2,800,000 GSF

④

Building Name : **Aon Center**
Address : 200 E. Randolph
Year Built : 1972
Architect : Edward Durell Stone
# of Stories : 83 Stories
Occupancy : Office
Area : 2,700,000 GSF

⑤

Building Name : **AT&T - USG Complex**
Address : 227 W. Monroe
Year Built : 1986-1989
Architect : Adrian Smith - SOM
# of Stories : 35 Stories
Occupancy : Office
Area : 2,700,000 GSF

⑥

Building Name : **Chase Plaza**
Address : 10 S. Dearborn
Year Built : 1969
Architect : Murphy & Assoc.
# of Stories : 65 Stories
Occupancy : Office
Area : 2,200,000 GSF

⑦

Building Name : **Two Prudential Plaza**
Address : 180 N. Stetson
Year Built : 1990
Architect : Loebl, Schlossman & Hackl
# of Stories : 64 Stories
Occupancy : Office
Area : 1,700,000 GSF

⑧

Building Name : **Three First National Plaza**
Address : 70 W. Madison
Year Built : 1981
Architect : SOM
# of Stories : 57 Stories
Occupancy : Office
Area : 1,400,000 GSF

⑨

Building Name : **Civic Opera**
Address : 20 N. Wacker
Year Built : 1929
Architect : Graham, Anderson, Probst
# of Stories : 45 Stories
Occupancy : Office/Opera House
Area : 1,200,000 GSF

⑩

Total Area: 25,900,000 GSF

EXAMPLE BUILDING DISTRICT BY BUILDING TYPE

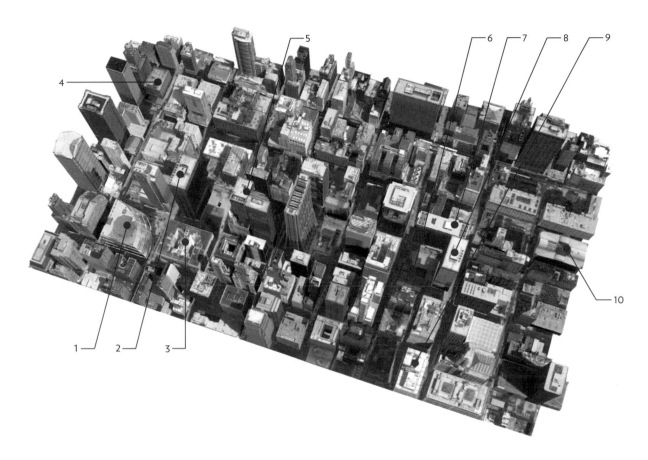

Building Name : **Thompson Center**
Address : 100 W. Randolph
Year Built : 1985
Architect : Helmut Jahn
# of Stories : 19 Stories
Occupancy : Office/Retail
Area : 1,557,654 GSF

1

Building Name : **Daley Center**
Address : 50 W. Washington
Year Built : 1965
Architect : Murphy & Naess
# of Stories : 32 Stories
Occupancy : Office
Area : 1,234,848 GSF

2

Building Name : **City Hall**
Address : 48 N. Clark
Year Built : 1911
Architect : Holabird & Roche
# of Stories : 12 Stories
Occupancy : Office
Area : 1,352,946 GSF

3

Building Name : **H. Washington College**
Address : 30 E. Lake
Year Built : 1962
Architect : Unknown
# of Stories : 11 Stories
Occupancy : School
Area : Unknown

4

Building Name : **Cook County Admin.**
Address : 69 W. Washington
Year Built : 1965
Architect : Jacques Brownson
# of Stories : 35 Stories
Occupancy : Office
Area : 1,044,938 GSF

5

Building Name : **Fed. Plaza Post Office**
Address : 230 S. Dearborn
Year Built : 1975
Architect : Mies van der Rohe
# of Stories : 1 Story
Occupancy : Post Office
Area : Unknown

6

Building Name : **Federal Plaza Bldg. 1**
Address : 219 S. Dearborn
Year Built : 1969
Architect : Mies van der Rohe
# of Stories : 39 Stories
Occupancy : Office
Area : 1,243,980 GSF

7

Building Name : **Federal Plaza Bldg. 2**
Address : 230 S. Dearborn
Year Built : 1975
Architect : Mies van der Rohe
# of Stories : 45 Stories
Occupancy : Office
Area : 1,333,800 GSF

8

Building Name : **Federal Reserve Bldg.**
Address : 230 S. LaSalle
Year Built : 1922/1960/1986
Architect : Graham, A.,P. & White
# of Stories : 14 Stories
Occupancy : Office
Area : 985,000 GSF

9

Building Name : **H. Washington Library**
Address : 400 S. State
Year Built : 1991
Architect : Hammond, Beeby & Babka
# of Stories : 8 Stories
Occupancy : Library
Area : 756,640 GSF

10

EXAMPLE ENERGY DISTRICT BY LOCATION AND TYPE

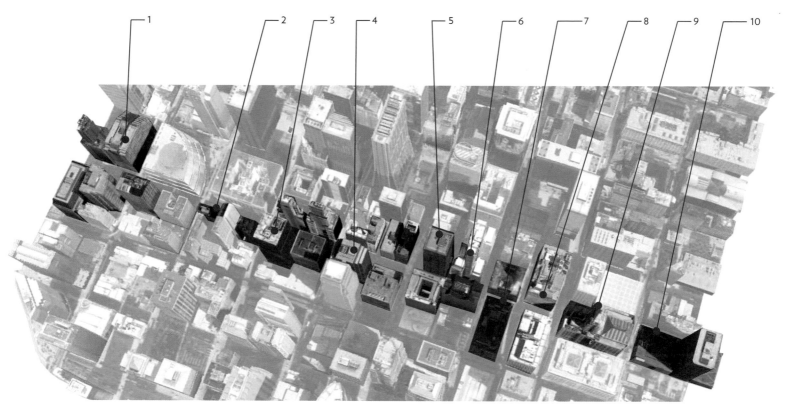

Building Name : **Loop Transit Center**
Address : 203 N. LaSalle
Year Built : 1985
Architect : SOM
# of Stories : 27 Stories
Occupancy : Office
Area : 775,800 GSF

1

Building Name : **Metropolitan**
Address : 134 N. LaSalle
Year Built : 1926
Architect : Rapp & Rapp
# of Stories : 22 Stories
Occupancy : Office
Area : 228,729

2

Building Name : **30 N. LaSalle Building**
Address : 30 N. LaSalle
Year Built : 1974
Architect : Welto Backet & Associates
# of Stories : 44 Stories
Occupancy : Office
Area : 1,106,765 GSF

3

Building Name : **10 S. LaSalle**
Address : 10 S. LaSalle
Year Built : 1986
Architect : Moriyama & Teshima
# of Stories : 37 Stories
Occupancy : Office
Area : 733,633 GSF

4

Building Name : **Harris Towers**
Address : 115 S. LaSalle /
　　　　　　　111 W. Monroe
Year Built : 1910/1958/1975
Architect : Shepley, Rutan/SOM
# of Stories : 23+15+38 Stories
Occupancy : Office
Area : 1,375,400 GSF

5

Building Name : **Bank of America Building**
Address : 135 S. LaSalle
Year Built : 1931
Architect : Graham, Anderson,
　　　　　　　Probst, and White
# of Stories : 45 Stories
Occupancy : Office
Area : 2,486,402 GSF

6

Building Name : **ABN AMRO Building**
Address : 208 S. LaSalle
Year Built : 1914
Architect : Daniel Burnham
# of Stories : 20 Stories
Occupancy : Office
Area : 1,041,763 GSF

7

Building Name : **Bank of America Bldg.**
Address : 231 S. LaSalle
Year Built : 1922
Architect : Graham, Anderson,
　　　　　　　Probst, and White
# of Stories : 36 Stories
Occupancy : Office
Area : 1,116,225 GSF

8

Building Name : **Chicago Board of Trade**
Address : 400 S. LaSalle
Year Built : 1930
Architect : Holabird & Root
# of Stories : 44 Stories
Occupancy : Comemrcial
Area : 2,132,988 GSF

9

Building Name : **Chicago Stock Exchange**
Address : 400 S. LaSalle
Year Built : 1985
Architect : SOM
# of Stories : 39 Stories
Occupancy : Office
Area : 1,100,000 GSF

10

Total Area:　12,097,706 GSF

EXAMPLE ENERGY DISTRICT BY STREET

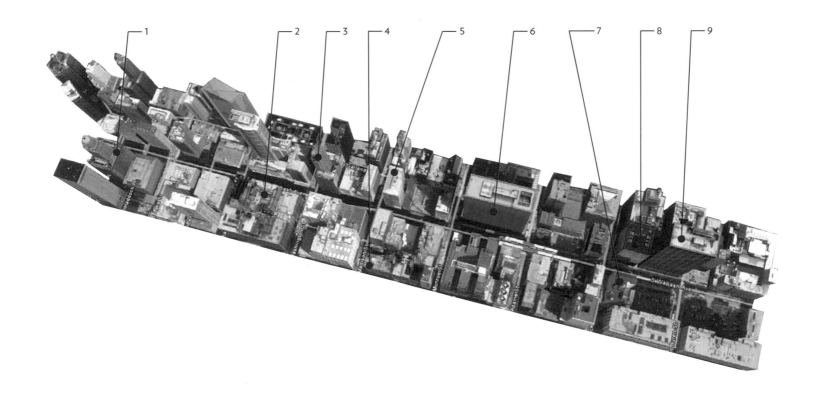

Building Name : **Jewelers Building**
Address : 35 E. Wacker
Year Built : 1927
Architect : Thielbar & Frugard
# of Stories : 40 Stories
Occupancy : Office
Owner : Unknown
Area : 966,720 GSF

❶

Building Name : **Marshall Fields**
Address : 122 N. Wabash
Year Built : 1914
Architect : Graham, Burnham & Co.
# of Stories : 12 Stories
Occupancy : Department Store

Area : 1,600,000 GSF

❷

Building Name : **Pittsfield Building**
Address : 55 E. Washington
Year Built : 1926-1927
Architect : Graham, Anderson, Probst, and White
# of Stories : 38 Stories
Occupancy : Office
Area : 515,280 GSF

❸

Building Name : **Carson Pirie Scott**
Address : 5 S. Wabash
Year Built : 1899-1961
Architect : Sullivan, Burnham
# of Stories : 12 Stories
Occupancy : Office/Dept. Store
Area : 110,640 GSF

❹

Building Name : **Mallers Building**
Address : 5 S. Wabash
Year Built : 1910
Architect : Christian Eckstrom
# of Stories : 21 Stories
Occupancy : Office
Area : 352,296 GSF

❺

Building Name : **Mid Continental Plaza**
Address : 55 E. Monroe
Year Built : 1972
Architect : Goettsch Parners Inc.
# of Stories : 49 Stories
Occupancy : Office
Area : 2,899,394 GSF

❻

Building Name : **DePaul College of Law**
Address : 25 E. Jackson
Year Built : 1917
Architect : Graham, Burnham & Co.
# of Stories : 16 Stories
Occupancy : School
Area : 540,000 GSF

❼

Building Name : **CNA Plaza North**
Address : 55 E. Jackson
Year Built : 1962
Architect : Naess & Murphy
# of Stories : 24 Stories
Occupancy : Office
Area : 522,140 GSF

❽

Building Name : **CNA Plaza Building**
Address : 333 S. Wabash
Year Built : 1973
Architect : Graham, Anderson, Probst, and White
# of Stories : 44 Stories
Occupancy : Office
Area : 1,300,000 GSF

❾

Total Area: 8,806,470 GSF

PILOT BUILDINGS SURVEY PROGRAM

Achieving increased energy efficiency in existing buildings has three necessary steps: auditing, retrofitting and commissioning. The Pilot Buildings Survey Program is the first step on this path.

According to the Energy Information Administration 2003 Commercial Buildings Energy Consumption Survey, office buildings have more floor area (12.2 billion square feet) than any other building type in the U.S., and have the highest total energy consumption (1.1 quadrillion Btu) of any building type. The largest buildings have a higher energy use intensity than any other size of building. This scale creates many opportunities for energy savings. A good place to begin is with an energy audit of the building.

Via a letter from the City of Chicago, about 80 large buildings in the study area were invited to participate in a survey through their respective property management firms. The buildings were chosen because of their large floor area or their inclusion as part of a larger portfolio of buildings owned or operated by a major owner or real estate management company. This group of buildings accounted for 73% of the total floor area in the study.

A questionnaire, with topics ranging from programmatic uses to mechanical, electrical and plumbing systems, was sent to respondents to be filled out before the interview. The questionnaire

was reviewed by Adrian Smith + Gordon Gill Architecture and Environmental Systems Design engineers before setting up an interview with the building manager and chief engineer.

The interviews generally lasted two hours, with the first hour spent discussing the questionnaire responses and the second hour spent performing a walk-through audit with the building engineer. The interview and walk-through process revealed that there are many lessons to be learned beyond simple efficiency strategies. Interviewing the engineers about their overall strategy and philosophy for operating the building in an efficient way was very informative. Although their pride and intimate knowledge of the buildings were impressive, some of the engineers had operated their buildings for decades without any focus on potential operational changes to improve efficiency. The highest-performing buildings in the survey were generally operated by engineers who had experience with a range of building types and systems. We also spoke about the reasons why building managers and owners do not undertake efficiency upgrades. Funding issues aside, tenant satisfaction was at the top of the list. If large-scale efficiency upgrades would potentially disturb a tenant, the work was not usually undertaken. A building manager's ultimate goal is to keep tenants in the building, which often conflicts with energy-efficiency efforts.

Ultimately, 13 buildings participated in the first phase of the Pilot Buildings Survey Program. These 13 buildings account for nearly 20 million square feet of Loop office space. Participant buildings were given a four-page summary that included a blind comparison of their building against the average survey building. Also included was a Carbon Reduction Recommendations Matrix, which describes long- and short-term, low-cost and capital-intensive improvements that can be made to each building. Participant buildings will be kept abreast of funding initiatives currently underway that can support the implementation of the

suggestions presented in the recommendations matrix.

The data collected in each building survey will be added to the data model described elsewhere. The water use, recycling and occupancy data will help to make possible the calculation of an overall carbon generation number for each building.

Carbon Reduction Recommendations Matrix

System		Current Operations	Low/No Cost Carbon Reduction Projects	End of Life Replacement/ Capital Intensive Projects
Energy	Fan Systems	**High rise system:** Four supply and return fans with VFDs. All but one supply fan has been upgraded to high efficiency motor. **Low rise system:** Four supply and return fans with variable pitch control. No upgrades.	1. Morning warm-up/ cool down: set OA damper position to 0% prior to occupancy. 2. Reduce lobby operating hours to better match peak occupancy. 3. Instruct tenants and/or cleaning staff to pull blinds at night, especially on eastern exposures in the summer. 4. Install occupancy sensors in the remaining base building bathrooms.	1. Consider installing a high efficiency pony chiller and associated pumps with VFDs to handle tenant loads. 2. Install VFDs on chillers, CHW and CW pumps, and cooling towers. 3. Fix variable pitch fans and install high efficiency motors with VFDs. 4. Investigate green roof installation options. Determine if current roof can handle additional weight loading.
	Cooling	Chilled water is supplied by two 1225 Ton, R-500 chillers. The chillers, CHW and CW pumps do not have VFDs. All 6 cooling tower fans do not have VFDs.		
	Heating	Electric reheat at terminal VAV boxes. Building controls temperature set points.		
	Domestic Water	Supplied via VFD pumps.		
	Occupied Hours	**Tower:** 8am – 6pm Mon.-Fri., 8am – 1pm Sat. **Lobby:** 24/7 Mon.-Sat., OFF Sun.		
	Lighting – Base building	All T-12's in common are in the process of being retrofitted with T-8's. Lighting is controlled by BAS; tenant after hours use is by request.		
	BAS	Digital control of air handling systems and monitoring of chillers, and terminal boxes. Installed in 2002. Digital controls for VAV's installed in 2008.		
	Envelope	Reflective Insulated Glazing units and insulated panel curtain wall.	None.	None.
Water	Fixtures	Original base building fixtures are not Energy Policy Act compliant.	None.	Upgrade all urinals, toilets, and faucets to high efficiency fixtures.
	Cooling Towers	Make up water is separately metered. Tenant condenser water is separately metered.	None.	Investigate plumbing restriction for installing a tenant cooling tower.
Tenant	Waste	Recyclables sorted on site. Recycle rate is above 50%.	Perform a waste audit to improve recycling rate.	None.
	Energy Efficiencies	Tenant newsletters and education.	Create lighting and power standards for future tenants that comply with the latest version of ASHRAE 90.1.	None.

A Carbon Reduction Recommendations Matrix (building information redacted for confidentiality)

PILOT BUILDINGS SURVEY PROGRAM
Pilot Building Participant Carbon Reduction Report

General Information

Building:
Year constructed:
Total floor area:
Number of stories:
Occupancy:
Population:

Category	[Redacted]	Average Program Building	Program Percentile
Energy (lcBtu/SF/yr)	60	90	100%
Water (gallons/person.yr)	10,432	6,955	10%
Recycling Rate (%/yr)	60%	44%	90%
CO2 Equivalent (kg/SF/yr)	13	19	90%

Energy

[Redacted] is an all-electric building. Review of electrical utility data show energy use peaks during the major heating months because electrical elements at the perimeter VAV boxes are commanded ON within tenant space. During the major cooling months, electricity use peaks because chillers, cooling tower and supply fans are running close to 100% during occupied hours. As can be seen in Figure 1, keeping the building warm requires more energy than keeping it cool; therefore, energy and carbon reduction projects should focus on reducing energy use during winter months. Also, as can be seen in Figure 1, [redacted] monthly energy usage intensity (EUI) is considerably less than the average building's EUI.

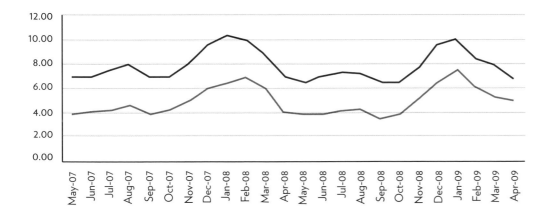

A portion of the four-page building report delivered to each building that participated in the survey

Buildings invited to participate in the Pilot Buildings Survey Program:

Building Address	Building Name	Owner	Manager
130 E Randolph	One Prudential Plaza	Bentley Forbes	Bentley Forbes
180 N Stetson	Two Prudential Plaza	Bentley Forbes	Bentley Forbes
190 S LaSalle	190 S LaSalle		CBRE
200 S Wacker	200 S Wacker		CBRE
29 N Wacker	Intertec Publishing		CBRE
55 E Monroe	Mid Continental Plaza		CBRE
50 S LaSalle	Chicago Board of Trade	CME	CBT Corp
111 W Jackson	TransUnion Building	Marmon Holdings	CBT Corp
200 N Columbus	Fairmont Hotel	Fairmont Raffles Hotels	Fairmont Raffle
33 N LaSalle	American National Bank		Golub
1 N Wacker	UBS Tower	D B Group	Hines
111 W Monroe	Harris Bank 1	Harris Bank 1	Hines
115 S LaSalle	Harris Bank 2	Harris Bank 2	Hines
1 S Dearborn	1 S Dearborn		Hines
191 N Wacker	191 N Wacker		Hines
70 W Madison	Three First National Plaza		Hines
151 E Wacker	Hyatt Regency Hotel	Hyatt	Hyatt
200 W Jackson	Bank of America		B of A
231 S LaSalle	Bank of America		B of A
135 S LaSalle	Bank of America Building		B of A
10 S Dearborn	Chase Plaza	Chase	JLL
30 S Wacker	Chicago Mercantile Exchange	CME	JLL
440 S LaSalle	One Financial Place	LaSalle Club Hotel	JLL
20 N Wacker	Civic Opera Building	Lyric Opera	JLL
333 W Wacker	333 W Wacker	Nuveen	JLL
1 S Wacker	One South Wacker	TIAA CREF	JLL
225 W Randolph	AT&T		JLL
79 W Monroe	Bell Federal Building		JLL
22 W Washington	CBS		JLL
150 S Wacker	Charles Schwab		JLL
230 S LaSalle	Federal Reserve Building		JLL
101 N Wacker	Hartmarx		JLL
19 E Madison	Heyworth Building		JLL
175 W Jackson	Insurance Exchange		JLL
303 E Wacker	KPMG Building		JLL
122 S Michigan	National Louis University		JLL
301 W Madison	Northwest Mutual Life		JLL
125 S Wacker	Northern Trust Bank		JLL
150 N Michigan	Smurfit Stone Building		JLL
228 S Wabash	Starck Building		JLL
20 S Clark	Two First National Plaza		JLL
77 W Wacker	United Building		JLL
1 E Wacker	Unitrin		JLL
55 W Monroe	Xerox Center		JLL
19 N Dearborn	19 N Dearborn		JLL
19 S LaSalle	19 S LaSalle		JLL
33 W Monroe	33 W Monroe		JLL
1 N State	Mandel Brothers Store		JLL
225 W Wacker	225 W Wacker		John Buck
325 S Wabash	CNA Plaza		John Buck
35 W Wacker	Leo Burnett Building		John Buck
209 S LaSalle	The Rookery		John Buck
111 S Wacker	111 S Wacker		MB Realty
311 W Monroe	311 W Monroe	Harris Trust	MB Realty
225 N Michigan	Michigan Plaza North		MB Realty
205 N Michigan	Michigan Plaza South		MB Realty
1 N LaSalle	One North LaSalle		MB Realty
1 N Dearborn	State & Madison Building		MB Realty
200 W Jackson	200 W Jackson		MB Realty
181 W Madison	181 W Madison		MB Realty
200 W Monroe	200 W Monroe		MB Realty
230 W Monroe	230 W Monroe		MB Realty
111 E Wacker	One Illinois Center	Parkway	Parkway
303 E Wacker	Three Illinois Center	Parkway	Parkway
233 N Michigan	Two Illinois Center	Parkway	Parkway
200 E Randolph	Aon Center	Piedmont Realty Trust	Piedmont Re
131 S Dearborn	Citadel Center		Prime Realty
323 E Wacker	Swiss Hotel	Swiss Hotels & Resorts	Swiss Hotels
200 W Madison	Madison Plaza	Hyatt	Tishman Spey
161 N Clark	Chicago Title & Trust		Tishman Spey
222 N LaSalle	Merrill Lynch		Tishman Spey
1 N Franklin	USG Building		Tishman Spey
227 W Monroe	AT&T Corporate Center		Tishman Spey
30 N LaSalle	30 N LaSalle		Tishman Spey
205 S Wacker	Sears Tower		Tishman Spey
208 S Lasalle	ABN AMRO Building		Tishman Spey
205 W Wacker	Engineering Building		Tishman Spey
2 N Lasalle	2 N LaSalle		Tishman Spey
11 N Wabash	11 N Wabash		Tishman Spey
200 S Michigan	Borg Warner Building		Tishman Spey
25 E Washington	25 E Washington		Tishman Spey
27 E Monroe	Palmer House Office Center		Tishman Spey
150 N Wacker	150 N Wacker		Tishman Spey

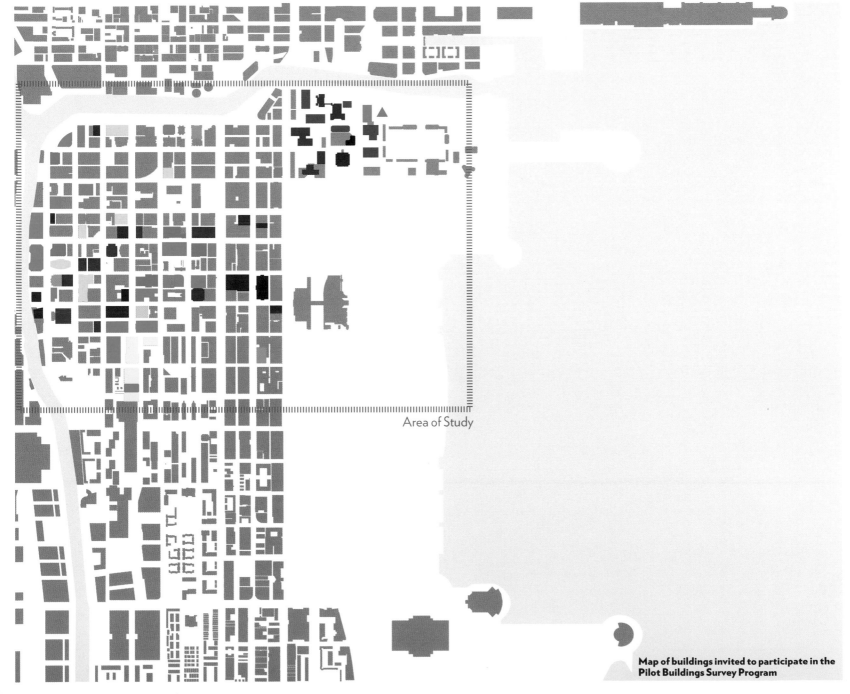

Area of Study

**Map of buildings invited to participate in the
Pilot Buildings Survey Program**

URBAN
MATRIX

A matrix is a point of origin, the place from which something else develops. The Loop is a matrix, and as such should reflect the city's ideals for the future: appreciation of diverse culture, dedication to family life and recreation, commitment to sustainability. The Loop must redefine itself as more than simply the city's center for commerce; it must transform into a vibrant, 24-hour district and a model for the accessible 21st-century community.

Carbon Reduction Strategies

What if the Loop became more than just a place for Chicagoans to work? What if, instead, it became increasingly accessible, green and family-friendly—the next great neighborhood for Chicagoans to live? Commercial space in the Loop area accounts for a majority of its land use and 90% of its carbon emissions. Energy loads in downtown office buildings are high, and the emissions number soars when you calculate the additional carbon output of every person who commutes downtown five days a week. By contrast, the Loop contains almost no residential space, and virtually none of the amenities and support infrastructure that many homeowners value in their neighborhoods, such as schools, daycare centers, parks and grocery stores. That forces those who work in the area to look elsewhere in the city for suitable housing and living environments.

In this chapter, we confirm a lower carbon footprint for mixed-use, 24/7 communities, which encourage higher-density living and more sustainable forms of transit.

The **Strategies** section examines a number of approaches to creating a vibrant urban core in the downtown area. We propose increased amenities and new schools within mini-neighborhoods in the overall Loop. We also examine calculations for the carbon emissions per square foot and per capita for a range of different percentages of commercial and residential areas.

We identify a number of **Precedents** of new residential buildings that have been converted from Class B and C office buildings in downtown Chicago.

Finally, a series of **Design Solutions and Pilot Projects** illustrate the conversion of identified existing buildings into new residential buildings and new high-performance commercial towers. A new parkway along Monroe, along with additional green roofs, new parks and permeable paving, will increase the Loop's green space. We also propose additional amenities such as new schools.

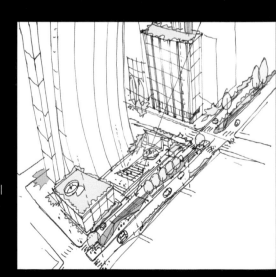

CARBON REDUCTION GOALS

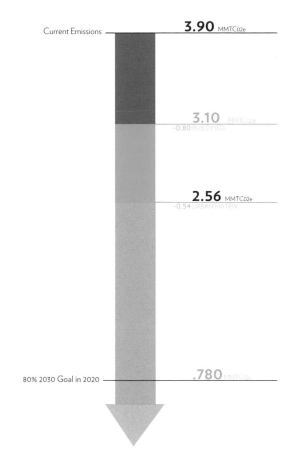

Current Emissions — **3.90** MMTC02e

3.10 MMTC02e
-0.80 BUILDINGS

2.56 MMTC02e
-0.54 URBAN MATRIX

80% 2030 Goal in 2020 — **.780** MMTC02e

Specific goals as related to building emission reduction within the DeCarbonization Plan

By increasing the ratio of residential to commercial office square footage, we can reduce a large portion of the carbon emissions due to the resulting decrease in commuters.

Further reductions in carbon emissions will come from the transfer of use from commercial to residential. This displaced office space will be superseded by high-performance, high-density green buildings in keeping with the DeCarbonization Plan.

Introducing permeable surfaces throughout the Loop in the form of pocket parks, permeable alleys and green roofs will mitigate the urban heat island effect. A reduction in roadway widths to accommodate green passageways will reduce the carbon impact within the Loop.

For these strategies, the total savings are estimated at .54 MMTCO2e.

Commercial use: 36kW/h

Residential use: 10kW/h

Green roofs can reduce noise pollution by 40 dB

Smart reflective roofs can reflect more than 80% of solar rays

Green roofs can cut mechanical loads by 30%

Transportation energy use per employee: 27,700 KBtu/yr

Average U.S. commute: 12.2 MI (one-way)

RELATIVE CARBON FOOTPRINT BY PROGRAM

The high volume of carbon emissions in the Loop study area can be partly traced to the area's density and the large proportion of square footage to land area. The area's primary function as a commercial district contributes to emissions as well. In general, commercial buildings have higher energy loads, consume more resources and produce larger volumes of waste than their residential neighbors. But the carbon footprint of a commercial building also absorbs the carbon footprints of the occupants who travel to it. When many of the occupants are traveling from other city neighborhoods or the suburbs, the emissions number increases significantly.

The amount of carbon contributed by commuters in their daily trips to and from the Loop significantly increases the overall carbon footprint of the study area and of the commercial buildings within it. In addition, the "suburb-commute" system is also inefficient.

A commuter living an hour's drive from work annually spends the equivalent of 12 work weeks or 500 hours in a car. According to a 2007 Department of Transportation report, traffic delays caused the United States to lose 3.7 billion hours and 2.3 billion gallons of fuel in 2003.

By contrast, those who live near work are far more likely to ride mass transit, bike or walk to work. An April 2007 study conducted by New York City's

Office of Long-Term Planning and Sustainability noted that, per-capita, carbon emissions of New Yorkers are less than one-third of those of the average American, due in part to more efficient, shorter commutes in New York City.

Dense, mixed-use, 24-hour neighborhoods take advantage of more sustainable methods of transit and reduce land use, conserving our most precious resources and contributing to the sustainability of our cities.

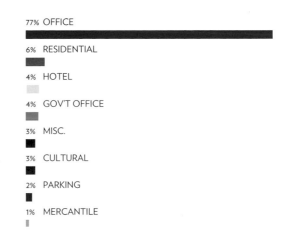

77% OFFICE

6% RESIDENTIAL

4% HOTEL

4% GOV'T OFFICE

3% MISC.

3% CULTURAL

2% PARKING

1% MERCANTILE

The "suburb-commute" method vs. the dense, mixed-use city. The advent of the 24-hour downtown community signifies the advent of a more sustainable era in urban living.

CO$_2$

SUBURBAN SPRAWL

17,000 kWh/person/year

MIXED-USE CITY

5,000 kWh/person/year

A VIBRANT URBAN CORE

The addition of residential real estate in downtown Chicago would lower the area's overall carbon footprint by creating a dense, mixed-use community. But adding significant residential space magnifies the area's need for amenities, which the Loop currently lacks, that support residential life: K-12 schools, daycare centers and grocery stores. A vibrant urban community relies on the quality of its amenities; therefore, the incorporation of these key elements is a critical step in the Loop's transformation into Chicago's next premier neighborhood for families.

The quality of renewed and additional amenities in the Loop is paramount. Strong schools will attract families and ensure the development of the community. Such amenities will need to grow as the mixed-use city evolves. Close proximity of amenities will offer extensive interaction within families. Sustainable grocers, daycare centers, schools, offices, retail and entertainment within blocks of one another will nearly eliminate private vehicle use within the Loop. The live-work environment will also require improvements to sustainable transit, an increase in public green space, district cooling and the reduction of stormwater treatment. These practices will result in enhanced air quality, reductions in the urban heat island effect and lighter mechanical loads.

The creation of neighborhoods in the Loop will need to ensure that amenities such as schools, grocery stores and parks are all within a few minutes' walk. ▶

2-3 min walk

© iStock/Christopher Futcher

The current occupancy breakdown in the study area is about 90% commercial and 10% residential. This amount of residential area, in addition to the amenities necessary to support that population (such as schools and grocers), falls short of the area that would be needed for even half of the daytime workforce.

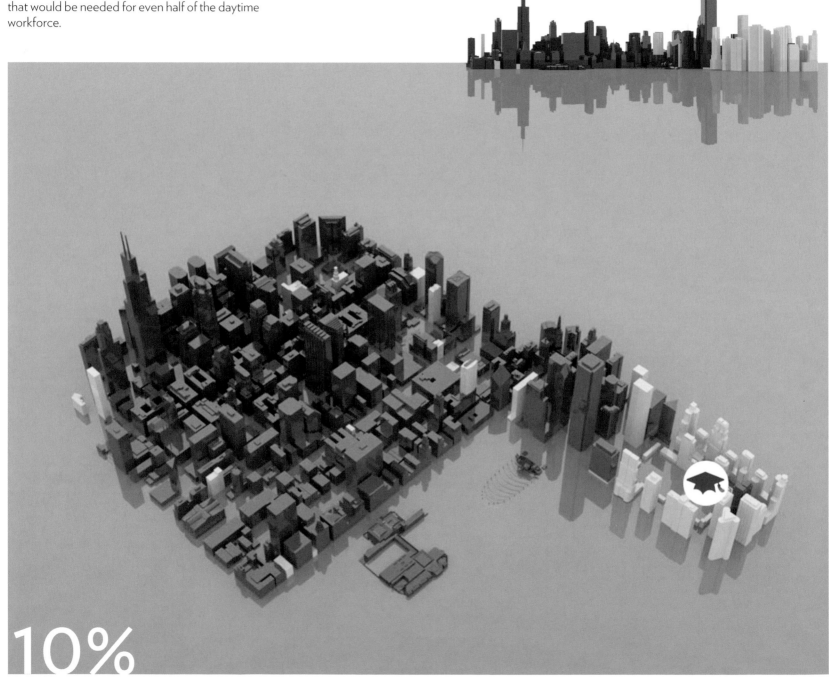

10%

An increase of residential space to 25%, with office space keeping pace with planned development, would necessitate the addition of at least one school, more amenities and more green space. Adding residential space and civic amenities could be accomplished by converting older, underperforming commercial space into residential units and transforming other existing commercial buildings into mixed-use spaces.

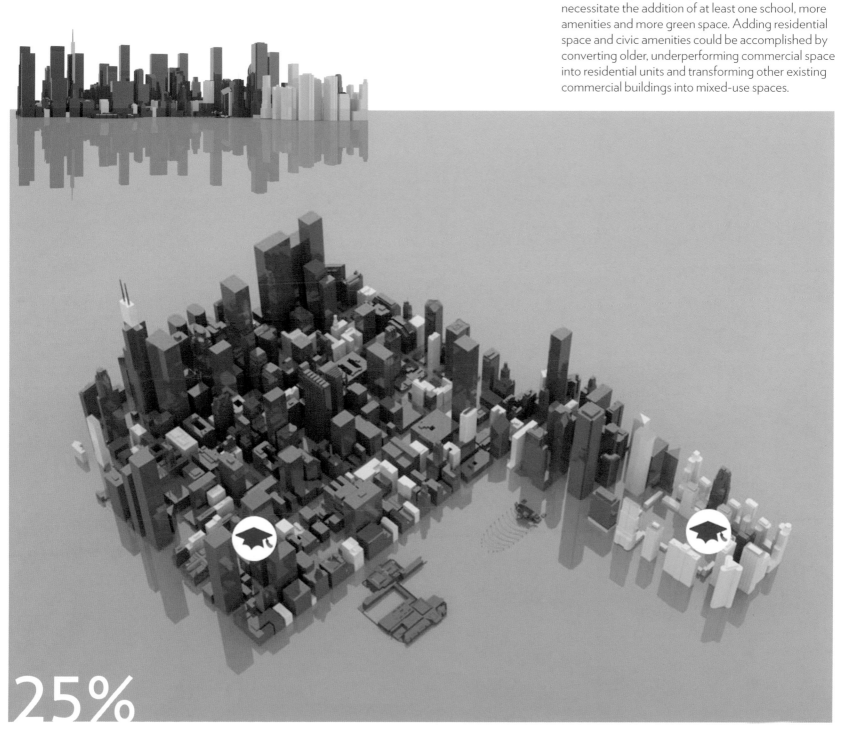

25%

At 40% residential, the area begins to look and feel like a truly mixed-use, dense community. The increased floor area would continue the pattern of outdated office conversion to residential use, with high-performance, newly constructed buildings making up the commercial shortfall to achieve planned office expansion. Residential amenities would continue to grow as required.

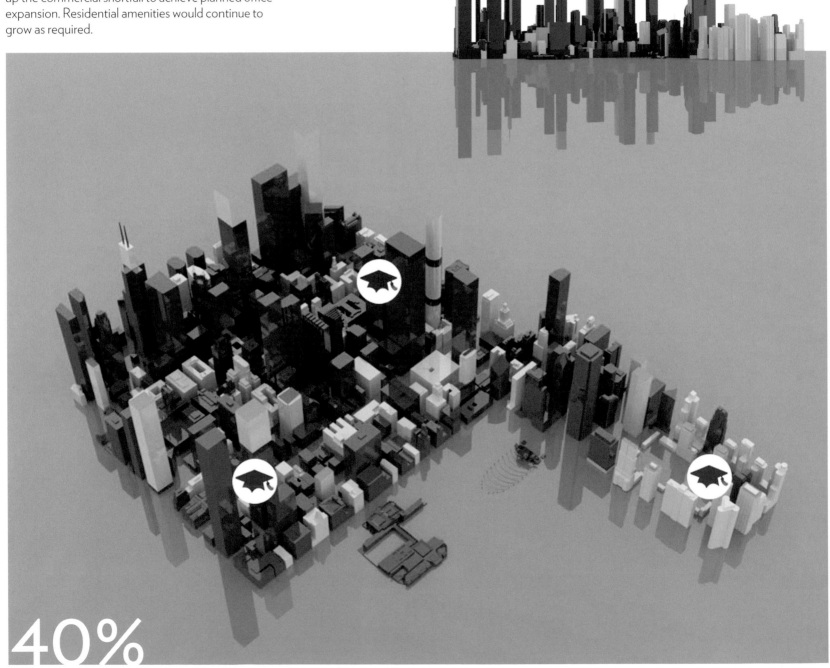

40%

As a result of carbon reductions from the decrease in commuting, a 50/50 commercial/residential occupancy mix in the Loop would result in one of the world's most efficient and sustainable live-work communities, creating a new paradigm for cities across the globe.

50%

REPLACE CLASS B AND C OFFICE WITH RESIDENTIAL

Current class B and C office space has a high vacancy rate and a low per-square-foot lease rate. According to the Appraisal Research Counselors' 2009 report, the central business district is made up of 51% Class B and C office space. Within the Loop, Class B and C tend to be older buildings, often with narrower floor plates and operable windows as well as higher vacancy rates. These buildings are not as leasable because they've become outdated. They may have lower floor-to-ceiling heights, lack the most up-to-date electrical and information systems infrastructure or may be difficult to access from public transportation. While there is no official or international definition of building classes, buildings are rated in relation to their neighboring community. Key elements in determining class are (from the Building Owners and Managers Association):

· HVAC capacity
· Elevator quantity and speed
· Backup power
· Security and life safety infrastructure
· Ceiling heights
· Floor load capacity
· Location

· Access (freeway, public transportation)
· Parking
· Construction, common area improvements
· Nearby and/or on-site amenities (drycleaning, restaurants, ATM, etc.)

Narrow floor plates and operable windows lend themselves well to residential conversion, ensuring that residents have plenty of access to natural light and air. The light and ventilation requirements of the Chicago Building Code require that all living spaces, including bedrooms, have access to daylight and operable windows.

Many of these buildings that are situated throughout the Central Business District could be part of the initial phase of transition to residential. Conversion of these buildings requires high-quality amenities to support the inhabitants' lifestyles. Foremost among these amenities are schools, daycare centers and grocery stores. Current clients of Thermal Chicago (the utility company that supplies district cooling for the city) are about 19% residential. They have plans to add more stations in coming years. Residential buildings are a good application for district cooling because of the small MEP area requirement and low maintenance requirements.

Displacing existing office space by converting it into residential units requires a plan to replace that space with new offices to meet the city's projected growth goals. This square footage should be replaced with new high-performance, high-density buildings. These buildings could also be mixed-use: office and residential, along with amenities such as schools, grocery stores, health care and other services required to support urban living. There are currently successful examples of repurposed and mixed-use conversion in the Loop that will be discussed further in this chapter.

Office building classifications

· Class A: These buildings represent the highest quality buildings in their market. They are generally the best-looking buildings with the best construction and high-quality building infrastructure. Class A buildings are also well-located, with good access and professional management. As a result, they attract the highest-quality tenants and also command the highest rents.

· Class B: These buildings are the next notch down. Class B buildings are generally a little older but still have good quality management and tenants. Often, value-added investors target these buildings as investments since well-located Class B buildings can be returned to their Class A glory through renovations such as facade and common area improvements. Class B buildings should generally not be functionally obsolete and should be well maintained.

· Class C: The lowest classification of office building and space. These are older buildings (usually more than 20 years) located in less desirable areas and in need of extensive renovation. Architecturally, these buildings are the least desirable and building infrastructure and technology is outdated. As a result, Class C buildings have the lowest rental rates, take the longest time to lease and are often targeted as redevelopment opportunities.

MDA CITY APARTMENTS
63 E. LAKE

High-end rental units
- Marketed to short- and long-term rentals
- Originally the Medical Dental Arts Building built in 1926
- 100% residential
- Has many sustainable design features including an accessible green roof
- Amenities include club room, exercise room, conference room, laundry room, bike storage and high-speed Internet access.

THE PARK MONROE
55 E. MONROE

Mid-range condominium $300,000-2,700,000
- Marketed to professionals and empty nesters
- Mixed-use building—top 10 stories converted to residential from Class B office
- Built in 1972; conversion completed in 2009
- Glazing replaced with double-pane, low-E glass
- HVAC system converted to district cooling
- Amenities include spa, pool, media room and fitness center
- Unobstructed views of the lake and Grant Park

THE FISHER BUILDING
343 S. DEARBORN

Competitive Loop rental units
- Marketed to young professionals, empty nesters and graduate students
- Residential with retail and office space on first and second floors
- Built in 1896; example of the First Chicago School of Architecture; terra cotta facade
- Range of units from studios to three bedrooms
- Amenities include club room, exercise room, conference room, laundry room, bike storage and high-speed Internet access.

POTENTIAL REUSE

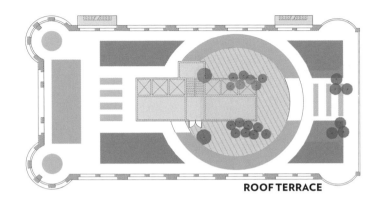

ROOF TERRACE

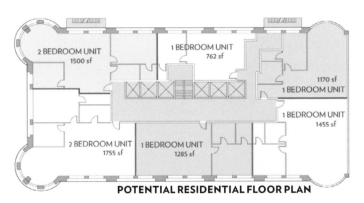

2 BEDROOM UNIT
1500 sf

1 BEDROOM UNIT
762 sf

1170 sf
1 BEDROOM UNIT

1 BEDROOM UNIT
1455 sf

2 BEDROOM UNIT
1755 sf

1 BEDROOM UNIT
1285 sf

POTENTIAL RESIDENTIAL FLOOR PLAN

Old Colony Building

EXISTING OFFICE

POTENTIAL LIVING

As the number of undesirable office buildings increases, so do opportunities for urban reuse. The central Loop is home to a large number of dated and landmarked buildings, many of which offer opportune spaces for residential occupancy. These buildings showcase narrow floor plates, operable windows, and are currently class B or C office spaces. Razing these structures can only contribute to rising global emissions and is not an efficient solution. With intelligent design, these buildings can be reclaimed, revitalizing the community and increasing property value while maintaining the city's heritage.

Hotel 71 was originally designed as a residential building, lending itself to adaptive redesign for residential conversion. The narrow floor plate and location are attractive features for residential space.

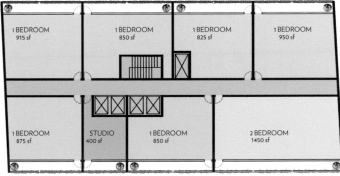

POTENTIAL RESIDENTIAL FLOOR PLAN

Addition of an accessible roof terrace will mitigate the Loop's urban heat island effect and offer a unique space with spectacular views for residents.

REPLACEMENT BUILDINGS

HIGH-PERFORMANCE BUILDINGS

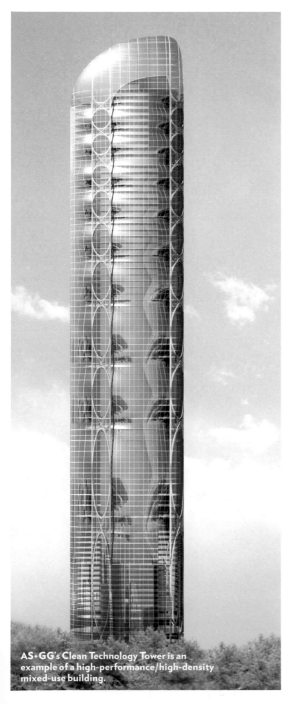

AS+GG's Clean Technology Tower is an example of a high-performance/high-density mixed-use building.

Buildings are universal. They shelter virtually every aspect of our lives—we live, work, learn, govern, heal, worship and play in them. As a result, buildings have a significant impact on energy use and the environment. According to the United States Energy Information Agency, homes and commercial buildings use 71% of the electricity in the U.S.; this number will rise to 75% by 2025. Opportunities abound for reducing the enormous amount of energy consumed by buildings.

High-performance buildings meet this challenge. A commercial building's design and ongoing monitoring are critical to achieving optimal energy, economic and environmental performance. Some of the key steps in designing an energy-efficient, net-zero-energy commercial building and assessing its performance are to use a holistic approach that considers all building components and systems during the design phase; to set specific, measurable goals for the project's energy use; and to develop an energy model of the building using simulation software, which helps in making critical decisions about a building's design early in the process. A holistic process for commercial building benchmark models can be used as a starting point with simulation software. The use of standardized performance metrics will help measure, report and evaluate energy performance.

In 1995, Four Times Square was the first speculative office building to be developed in New York City in almost a decade. It has 48 stories of occupied space and was the first project of its size and financial structure in the U.S. to adopt standards for energy efficiency, sustainable materials, and indoor environmental quality, as well as for responsible construction, operations and maintenance practices. Since then, enormous improvements in standards and technology have been made through the efforts of such organizations as the U.S. Department of Energy, the U.S. Green Building Council (USGBC) and the American Society of Heating, Refrigerating

and Air-Conditioning Engineers (ASHRAE). The U.S. Department of Energy's Net-Zero-Energy Commercial Building Initiative aims to achieve marketable net-zero-energy commercial buildings by 2025. Net-zero-energy buildings generate as much energy as they consume through efficiency technologies and on-site power generation. ASHRAE, an organization that has set standards for high-performance buildings, is soon to release its Proposed Standard 189.1, *Standard for the Design of High Performance, Green Buildings Except Low-Rise Residential Buildings*, in conjunction with the Illuminating Engineering Society (IES) and the USGBC. The standard is slated to be the first code-intended commercial green building standard in the U.S.

Funding

A number of the programs discussed in the Funding chapter are relevant to the Urban Matrix carbon reduction strategies. The federal programs encourage residential projects. For example, the EECBG discretionary grant application currently requires a mix of uses, including residential. The Tax Increment Financing program is well adapted for Class B and Class C building renovation.

One of the principal purposes of the LaSalle Central Tax Increment Financing's Redevelopment Project Plan was the renovation of this type of aging building stock. The plan recognizes that the building stock must be renovated to ensure a vibrant city. Another feature of the Illinois TIF law is the requirement of a set-aside of tax increment in a project that receives a TIF benefit if that project has a residential component. This feature of the Illinois TIF law should be explored to find ways in which it could be used in combination with other tools to assist in the development of schools serving the Loop. Another feature of the TIF law allows the use of tax increment for public buildings, including schools, as well as certain training and daycare expenditures.

Permeable strategies above grade

Permeable strategies at grade

PERMEABLE SURFACES

At grade

Pocket parks
Pocket parks offer green social and recreational spaces at ground level to support the eventual growth in the residential population within the Loop. They provide areas for residents to walk dogs and have lunch, and for children to play outside. These parks can also offer a location for special seasonal events and daily or weekly outdoor markets. They also increase the permeable area for stormwater runoff to seep into the ground, reducing the amount of water requiring treatment.

Streets and alleys
While green roofs absorb rainwater, the majority of rainwater actually hits the vertical faces of buildings and runs down into the streets and alleys. This stormwater runoff is then required to be transported and treated. Incorporating permeable systems in alleys will capture this runoff so that it's absorbed rather than sent into the sewer system.

Green walls
By adding vegetated screens on the vertical and roof canopy areas of buildings and parking structures, air pollution can be reduced. The plantings absorb the noxious emissions from vehicles and store or metabolize air pollutants.

Green pathways
Green pathways can be developed by decreasing the impermeable paved width of some arteries. This axis or pathway will encourage and promote pedestrian and bicycle activity and provide further amenity space for Loop inhabitants. This also enhances the concept of neighborhoods that support activity 24/7, accommodating business during the day and social activities in the evening. These pathways also reduce the amount of stormwater runoff requiring treatment and mitigate the urban heat island effect.

Above grade

Green roofs
Various green roof types are appropriate for incorporating into new and existing buildings. Extensive green roofs should be incorporated on all residential buildings so that the space can be inhabited and enjoyed by residents. These rooftops can also be used for play and as exercise areas for schools, or they can be incorporated into urban agriculture projects. Shallower green roofs can be used on other building types to reduce stormwater runoff and the urban heat island effect. They also add value to the views that residents see from their windows.

ENVIRONMENTAL LEARNING SCHOOL

A superior education system is vital to the evolution of the mixed-use city. Facilities must be state-of-the-art with advanced programs and a diverse curriculum. Sustainability issues such as genomics, biodiversity and agroforestry should be introduced to students early on. Students should be encouraged to engage with the integrated natural environment and experience hands-on learning.

Schools in the Loop will enjoy great proximity to the heart of Chicago, allowing for increased interaction with the community as well as nearby satellite schools. These interactions are dependent on the proximities created by the mixed-use community. The contingence of life in the Loop will also enhance interfamily interactions. With a diverse student body and a holistic approach to early development, institutions such as the Daley Environmental School will become prototypes for the Chicago Public Schools.

MENTAL SCHOOL

Monroe Street Green Corridor

Monroe Street Green Corridor

MOBILITY

The Loop has long been a hub for transit systems citywide; the area even takes its name from the form of the elevated tracks that surround it. As greener forms of transit re-emerge in the city, Chicago has an opportunity to create a sustainable infrastructure for the Loop that would serve as a model for efficient transit systems citywide.

An Interconnected Low-Carbon Transit Network

What if Chicago's existing grid could be augmented to create a sustainable network throughout the Loop—resulting in a transit system that's cost-efficient, energy-efficient and encourages healthy, low-carbon lifestyles? While many transit routes exist to bring commuters into and out of the general Loop area, the systems lack interconnectivity and some elements are outdated. To steer more of the population toward taking sustainable modes of transit, the city needs to augment existing transit and incorporate new systems to accommodate commuter needs.

In this chapter, we develop **Carbon Reduction Goals** from a combination of improvement to transit and pedestrian amenities.

In **Analysis**, the relationship between density and gas consumption is examined. Cities that have lower per capita gasoline consumption generally have dense, dynamic city centers. This analysis is valuable in determining goals for Chicago. We also analyze the existing transit infrastructure in Chicago, looking specifically at rail and bus commuting, bicycle routes, pedestrian routes and the existing taxi fleet.

In the **Strategies** section we explore logical, integrated solutions that also support the Urban Matrix concept. We study new trends and systems currently being explored in other areas of the United States and around the world.

Finally, we propose two significant **Design Solutions and Pilot Projects** that create mobility solutions. These projects, the Monroe Street

Intermodal Axis and Pedestrian and Bicycle Amenities, would enhance the quality of life of the residential and commercial users of the Loop.

CARBON REDUCTION GOALS

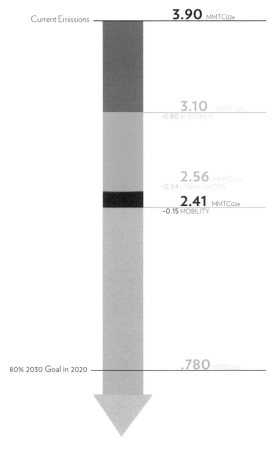

Current Emissions — **3.90** MMTCo2e

3.10 MMTCo2e
-0.80 BUILDINGS

2.56 MMTCo2e
-0.54 URBAN MATRIX

2.41 MMTCo2e
-0.15 MOBILITY

80% 2030 Goal in 2020 — **.780** MMTCo2e

Specific goals as related to transit emission reduction within the Loop De-Carbonization Plan

The DeCarbonization Plan study area is mainly comprised of the central business district of Chicago, and includes its energy-consuming buildings and the workers who commute to them. The strategies of this section are focused on increasing ridership of the existing and proposed upgraded mass transit systems into the Loop.

To make mass transit more appealing, we studied several issues. Cross-town access from the existing suburban transit hubs on the west side of the Loop is difficult. Offices in the River East neighborhood currently have to run shuttles from Metra stations to their buildings to keep their space competitive in the office real estate market.

While modes of transit are key in any discussion of transit-oriented development, the subject of walking is just as important. In a Chicago winter, encouraging walking requires sheltered paths to make the Loop a true transit-oriented neighborhood. Many experienced downtown commuter workers use buildings with through-block corridors to spend as much of

their foot commute as possible inside the buildings. There are existing, albeit ill-maintained, underground walkways in downtown Chicago to address this problem. There are also several grade-level through-block connections.

Bike paths are discontinuous, making it difficult to get from many parts of the city into the heart of the Loop. There are also significant safety concerns with existing Loop bike routes because they're shared with buses, taxis and cars rather than being dedicated solely for bicycles.

By updating to new, more efficient equipment, technologies and infrastructure for rail, bus and taxis, we can realize significant carbon savings. Improving pedestrian and bicycle amenities will also decrease automobile use. Adding new transit options such as one-way car/scooter/bicycle programs will further reduce the need for private car ownership, reducing carbon emissions.

Average daily CTA ridership is 1.6 million rides

Average daily Metra ridership is 300,000 rides

The Lakefront Bicycle Trail is 18 miles

Chicagoans can save $400/month by switching to public transportation

GASOLINE USE PER CAPITA

80,000 —

70,000 — ⬤ Houston

○ Phoenix

○ Detroit
60,000 — ○ Denver

⬤ Los Angeles

○ San Francisco
○ Boston
50,000 — ○ Washington DC

● Chicago

○ New York

40,000 —

○ Toronto

○ Perth
30,000 — ○ Brisbane
○ ○ Melbourne
Adelaide ⬤ Sydney

20,000 —

Hamburg
Stockholm ○ ○ Frankfurt
Paris ○ Zurich Brussels
London ○ Munich
Copenhagen ○ West Berlin
10,000 — Amsterdam ○ Vienna
Singapore ○ ○ Tokyo

DESIRED

○ Hong Kong

0 ○ Moscow

URBAN DENSITY

104

CHALLENGES

No convenient link from Metra and Pace to the rest of the Loop

Currently both Union Station and Ogilvie Station are disconnected across the river from the rest of the Loop. Walking, private shuttles and taxis are the only real options to get into the Loop. Many office buildings on the east side of the Loop lease private shuttle bus services to transport their tenants to and from work each day. Being close to public transportation is an important factor in the BOMA building classification system. Buildings without good access can be downgraded and thus receive less per square foot for their suites.

Safe bike paths, particularly running east–west, are limited in the Loop

Each year the city is adding to existing bike paths and lanes, and plans to continue to do this. Yet there is still no safe path to get from Union and Ogilvie or the West Loop to the east side of the Loop. Bicycle paths in the Loop area, where most cyclists need to commute to work, are shared with buses and cars, making them very susceptible to accidents.

The First Mile/Last Mile commuter issue

While there is public transit into the city, there are no expedient connections to get commuters from the stations to the West Loop, North Loop or South Loop. Surveys show that people will opt to drive if they have to walk more than 10 minutes to their final destination after their train commute.

No connected, safe, sheltered pedways during inclement weather conditions

There is a series of basements and walkways below grade in the Loop, but they are often disconnected, in poor repair or have unclear signage.

Bus system tracking could be improved

The CTA has implemented a Bus Tracker system that allows riders to know what time a bus will arrive. However, riders often find that when the bus arrives, it's full and drives past them. Bus traffic congestion is a common problem in the Loop.

Quality of El Train Cars and Stations

Upgrades to the El would improve rider comfort, convenience, safety and the quality of the rider experience to an international standard.

Taxi fleet is outdated

Most current taxis are re-purposed police cars that are not fuel-efficient. They can also be uncomfortable and contribute to air pollution.

This diagram plots the relationship between urban density and gasoline consumption. Colored circles show the number of Earths needed to sustain the lifestyles of each city.

© iStock/David H. Lewis

RAIL

Currently the Loop is well served by CTA El trains and buses. There are also four Metra stations within the study area. However, there are obvious abrupt disconnects between these systems. There are strategies in the current Chicago Central Area Action Plan that are designed to address this over time. These strategies include:

Carroll Avenue Transitway
A limited stop line is planned that will connect Union Station with Millennium Park Station along the north edge of the Chicago River, improving connections between the West Loop, River North and Streeterville.

Clinton Avenue Subway
A north–south subway is proposed through the West Loop, running from Chinatown and connecting to Ogilvie and Union Stations and running north to North and Clybourn.

East–West Transitway along Monroe
A below grade connection running east–west beneath Monroe, connecting Ogilvie and Union Stations Blue and Red El lines, is proposed. It would include access to the future connection to O'Hare Airport below Block 37 and terminate below grade at Millennium Park Station.

The above strategies would improve connectivity to the system.

The second main issue is the overall quality of the El ride. This can be improved by the following strategies:

Rider comfort
The international standard for most light rail is comfortable cushioned seats. Rail cars are also well-lit and spacious. Tracks and rolling stock should be well maintained to provide a smooth ride without excessive noise.

Rider information
The best transit systems, such as the London Underground, offer information to customers about when the next train will arrive. At stations with multiple train lines, the information includes the status of the different lines.

Transit stations
Transit stations should be well maintained as a representation of civic pride. Appropriate lighting and durable, high-quality finish materials are important.

Studies have shown that higher-quality rail transit does make a significant difference in ridership because riders feel safe and trains are convenient.

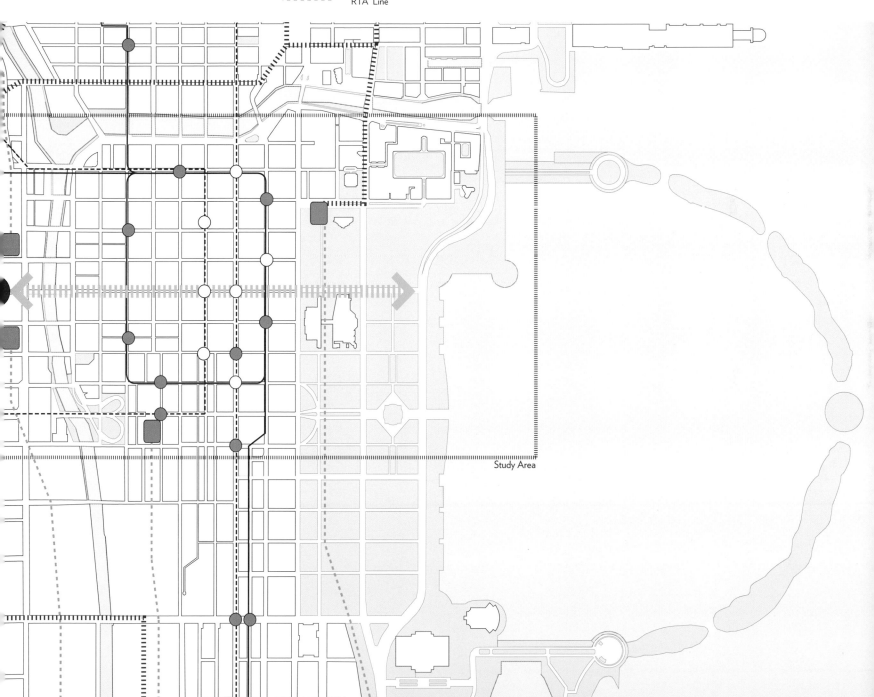

۱۱۱۱۱۱۱۱۱۱۱۱۱	Below Grade Transitway
۱۱۱۱۱۱۱۱۱۱۱۱۱	Clinton Subway
◄IIIIIII►	Monroe St Underground Intermodal Corridor
———	CTA Line (Above Ground)
- - - - -	CTA Line (Underground)
· - · - · - ·	RTA Line

▨	Existing Station
●	New Stop
●	Existing CTA Stop
○	Improved CTA Stop

Study Area

BICYCLE PATHS AND BICYCLE SHARING

Loop bicycle lane safety improvement

While neighborhood and lakefront pathways are well-marked, reasonably safe and well-used, the Loop is very difficult to navigate via bicycle. No dedicated bike lanes enter the central Loop. Loop streets that do have designated bike routes force cyclists to share the bus and auto lanes, which creates a potential for accidents. There's a strong need for at least one dedicated, safe east–west bicycle route from Canal Street to Millennium Park and a north–south bicycle route connecting north Loop and south Loop neighborhoods to the central Loop.

Bike repair, storage and showers

Safe storage, repair and access to showers promotes bicycle commuting. The theft rate for bicycles in the Loop is very high, even when bicycles are locked. There's currently one major bike station located in Millennium Park. Another such facility in the West Loop would be beneficial to commuters. It would also be beneficial to incentivize commercial buildings to provide bicycle storage and showers.

Riverwalk bike path

The Chicago Riverwalk forms an important east–west connection from the lakefront bicycle path to the central Loop. However, its bike path is not continuous and has the potential to be expanded.

Bicycle rentals

While there are bicycle rentals available at several locations in the Loop, they are focused on tourists rather than commuters. Their rates reflect this. No one-way bicycle share system is currently available. Rentals have the potential to be more commuter-friendly through better rates and one-way options. Buildings could also provide loaner bikes for daytime commuting within the Loop; one example is a program at Willis Tower in which tenants can borrow bicycles.

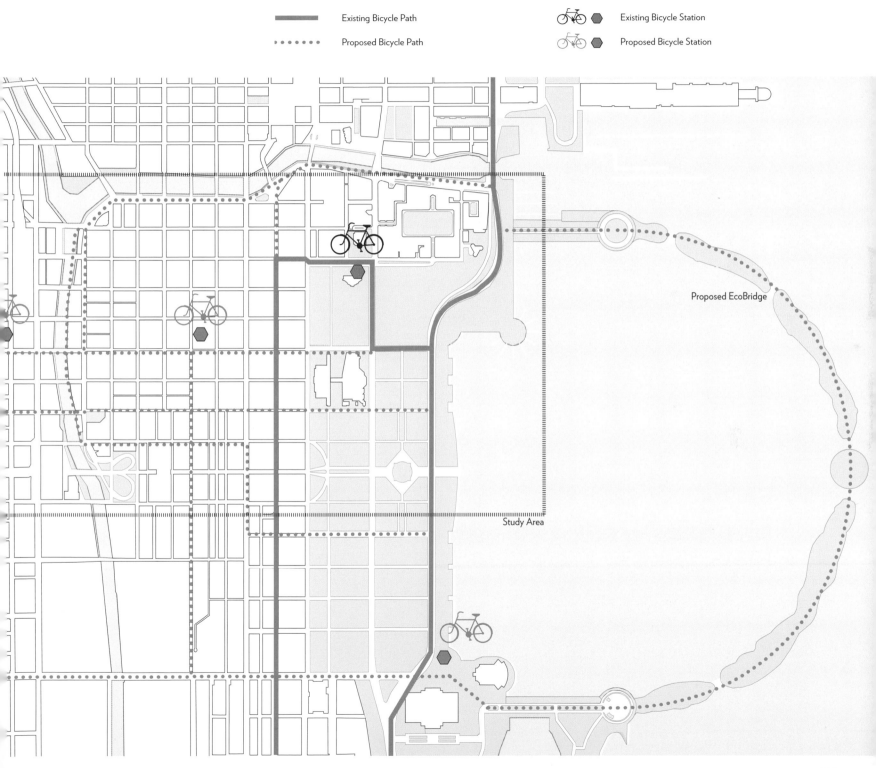

Existing Bicycle Path

Proposed Bicycle Path

Existing Bicycle Station

Proposed Bicycle Station

Proposed EcoBridge

Study Area

CAR SHARING AND SMART TRANSIT

Car sharing
There are currently many Zipcar and I-Go car-sharing locations sprinkled in convenient locations throughout the Loop. These services have one to three cars at each location. These are two-way car sharing services; the car must be returned to the location where it was picked up within a specific time that the member has stipulated.

There is currently no scooter sharing, bike sharing or one-way car sharing available in the Loop.

Parking space identification
A significant amount of traffic in the downtown area is the result of cars trolling for parking. With existing security monitoring infrastructure and additional remote sensing technology, the location and price of parking spaces could be transmitted to vehicles to expedite and improve the efficiency of this process.

Real-time congestion charging
Too many cars on too few roads—this is increasingly an issue for cities worldwide. A system in which vehicles are recognised and charged for roads use is a unique way to control the flow of traffic through economic motivators. Similar to the smart grid technology used for demand management of electricity, vehicles could be charged based on their emissions and the rush-hour congestion of certain roads.

On-street parking
Downtown Chicago has a significant amount of on-street parking, which contributes to traffic congestion through trolling cars and limits street width. More pedestrian-friendly streets with limited parking could relieve congestion. Due to the rising cost of street parking through the introduction of pay-boxes, the use of parking garages is a viable alternative for many people who visit or commute by car.

Truck traffic
Trucks currently comprise much of the traffic in the central Loop. Strategies discussed in the Waste chapter of this study discuss ways to reduce truck traffic due to waste disposal. Green construction practices, conservation of materials and material re-use can also limit the volume of deliveries needed in the Loop. Extra fees for trucks, limited hours for deliveries or stricter emissions standards could also reduce carbon emissions from trucks, but the economic impacts of such measures would need to be studied.

© Smart Cities Group/MIT Media Lab

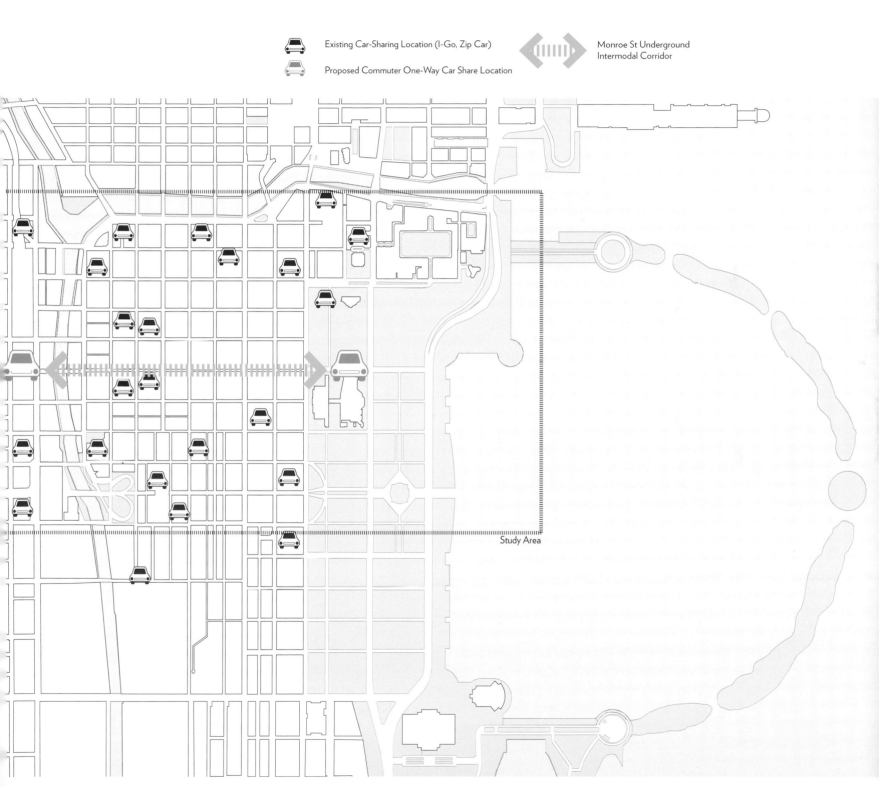

Existing Car-Sharing Location (I-Go, Zip Car)

Proposed Commuter One-Way Car Share Location

Monroe St Underground Intermodal Corridor

Study Area

TAXIS

Current taxi situation in Chicago

The majority of the current taxi fleet of 6,700 in Chicago consists of re-purposed police cars. These are neither fuel efficient nor comfortable. Of the 6,700 taxis only 50 are currently hybrids, based on the Chicago rule requiring taxi fleets with more than 50 vehicles to include a hybrid or alternative fuel car in their mix. Yet hybrid taxis can save taxi drivers up to 1,666 gallons of gas per 100,000 miles, reducing carbon emissions by 14.5MTCO2e per car per year.

Green taxis

A city requires a mix of energy efficient vehicle types to create an effective, low carbon solution to the taxi system. The current five-passenger vehicles are generally used to transport the driver and a single passenger. Not only is this a waste of energy, but the large vehicles also exacerbate congestion and make it more costly for drivers who purchase their own fuel—thus increasing the cost of fares in general. The most recent viable technology in taxis is the hybrid fuel cell taxi. The fuel cell provides electricity to the wheel-mounted electric motor. The taxis are able to run all day without refueling, have a maximum speed of 120 kilometers per hour and have faster braking than gas-fueled cars.

London fuel cell taxis

The current TX4 version is built by London Taxi International (LTI). LTI is partnering with Intelligent Energy, Lotus Engineering and TRW Conekt to build a test fleet of fuel cell powered London cabs. The fuel cell system provided by Intelligent Energy will fit into the engine compartment of the cab and will provide a maximum speed of 75mph and better acceleration than the standard diesel cab. Lotus will handle packaging and integration of the new electric drive system as well as the control systems to make it all work. TRW Conekt will handle the safety analysis, and testing of the controls, electrical systems and electronics. The goal is to have 50–100 fuel cell cabs operating on the streets of London before the 2012 Summer Olympic Games take place.

Minimodal by Hybrid Product Design

The Minimodal is a taxi design concept that was designed specifically for the city of New York. The vehicle design is small and light and uses a hybrid-power system. The two-person passenger cab is low to the ground and features side windows and a fully open skylight. The driver sits in front of an all-glass front end and on top of the hybrid engine. The car also has a signaling system that would let other drivers know when a passenger was coming out, to avoid collisions. The designers also envisioned two larger versions of the taxi to accommodate more passengers. The Maximodal would seat three and a wheelchair and the Mogulmodal would seat four plus a wheelchair.

Chicago Taxi prototype

Similar to the London cab model, the Chicago Taxi could be designed and built in Chicago, creating jobs and generating revenue that stays in the city. Partnering with a major American auto manufacturer with a local production plant, the construction of the cab could employ people in the factory, as well as beyond the confines of construction proper to local parts suppliers. The hybrid or fuel cell powered taxi prototype would be ruggedly designed, similar to checker cabs of the past, and include such amenities as a ramp and other universal design features to meet ADA requirements. It would also include global position indicating system screens so cab occupants could see their position or query other locations if they were not familiar with the city. The comfort of the prototype would be important, as the currently popular re-purposed police cars do not offer it in either finishes or ergonomics. Close attention would be given to the driver partition and a friendlier rear "social environment." The car would also include a signaling system to let other motorists know when a passenger was exiting the vehicle.

© Corbis/Ferdaus Shaman

INTELLIGENT ENERGY

© Corbis/Facundo Arrizabalaga/EPA

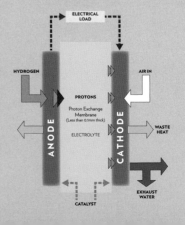

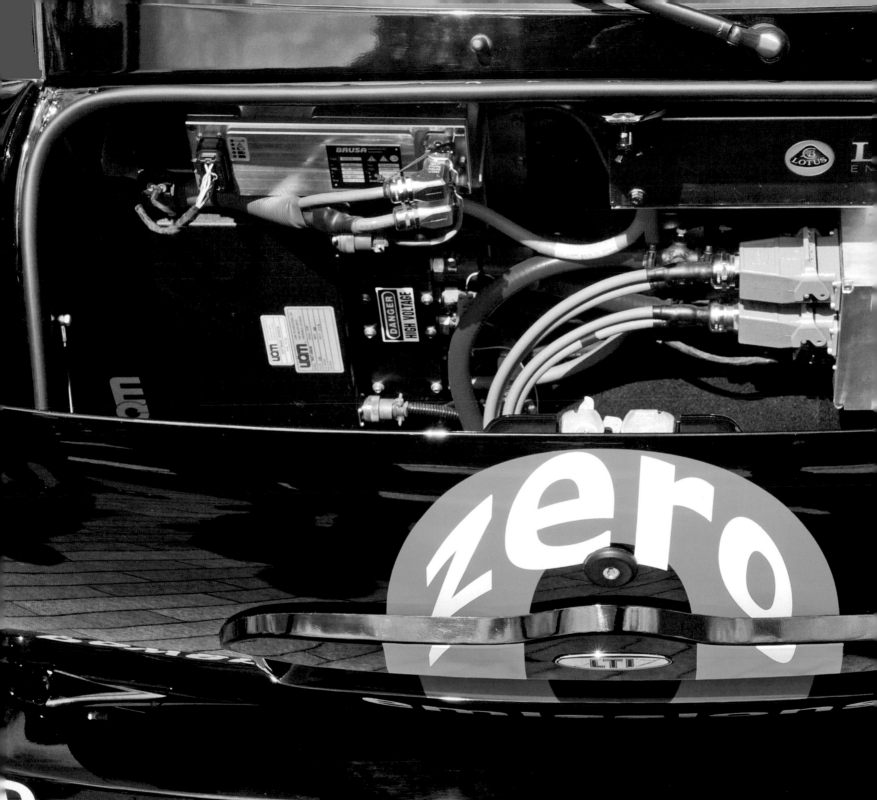

zero

DANGER
HIGH VOLTAGE

BRUSA

LOTUS

LTI

Powered by Hydrogen

UNDERGROUND PEDWAY

There is an extensive system of underground pathways below the Loop area. Most of these are disconnected, under-utilized and poorly maintained. There are sections that are neither pleasant nor safe to walk through. While there are numerous food courts in the basements of buildings, there are few higher quality amenities. There is no consistent connection and signage to transitways.

The underground pedway system has the potential to be revitalized if new policies are put into place. The first step is better connectivity. If a commuter knows that they will be able to reach their destination underground without coming to a dead end, new confidence in the system can be created. Connections to train stations and subway stations can also be incorporated for added convenience. An increase in pedestrian traffic will encourage retail activity below grade. Successful examples of this underground system are in the Canadian cities of Toronto and Montreal.

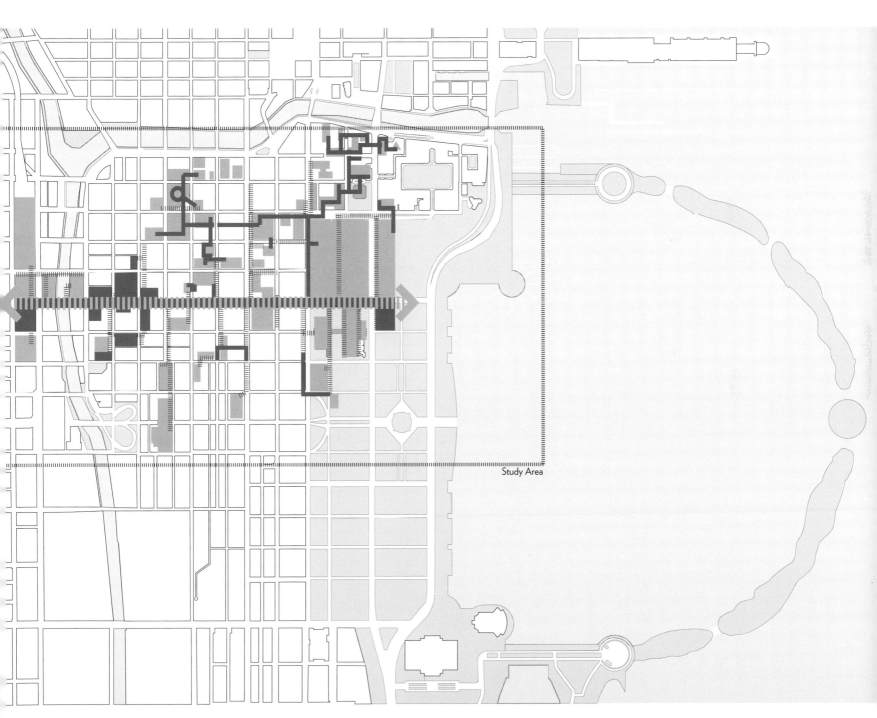

Existing Underground Amenities

Proposed Underground Amenities

Existing Underground Link

Proposed Underground Link

Monroe St Underground
Intermodal Corridor

Study Area

PEDESTRIAN INITIATIVES

Many cities are currently testing strategies to promote pedestrian mobility on streets that have traditionally been ruled by private cars and other forms of vehicular transportation.

These lane reduction strategies have been found to increase commercial activity. Many schemes can also incorporate recreational activity such as exercise classes and craft sessions for children.

Existing traffic is re-routed to reduce the traffic load in these commercial zones. There is also the potential to attract activity towards neighborhoods in need, improve air quality, reduce noise and increase city park area. While these strategies are generally temporary in nature—for example, Broadway in New York becomes a pedestrian zone at certain times—this shows a step in the right direction toward improving pedestrian networks in our cities.

New York and Los Angeles are two of many other cities that have experimented with temporary green spaces in place of street parking. These new urban park spaces prove to be quickly occupied and enjoyed by the public.

London has closed off Oxford Street to vehicular traffic on certain occasions to examine the toll this would take on traffic and retail purchases. Oxford Street has recently been permanently restricted to use by pedestrians, bicycles, public transit and taxis only.

1 **Times Square, New York**
2 **Pearl Street, New York**
3 **Park(ing) Day, Los Angeles**
4 **Oxford Street, London**

INTERMODAL TRANSIT

Integrating public transportation areas with access to train, bus, bike and taxi services, as well as linking to offices and residential areas, can create dynamic urban centers. These transit centers work to integrate above and below grade amenities, both existing and new.

Connections to airports by rail and regional rail systems make these centers hubs for urban activity.

5 Euralille Intermodal, Lille
6 Centraal Station, The Hague
7 Berlin Central Station, Berlin

ONE-WAY CAR AND BICYCLE SHARING

While car sharing is very popular in the Loop, it does not allow for one-way car sharing. One-way car sharing addresses the issue of First Mile/Last Mile commuters. These commuters could take public transportation but because their closest train is not within walking distance, they choose to drive to work. This can occur at the beginning or end of their trip. For example, a commuter arrives at Union Station but works in the River North neighborhood. One-way car sharing provides a system of small cars, bikes or scooters that can be collected at the train station and dropped off at a convenient location close to the final destination.

One such system is "Mobility on Demand." This is a system currently under development that provides stacks and racks of light electric vehicles and bicycles at closely spaced intervals throughout the city. When a commuter wants to go somewhere, he or she can simply walk to the nearest stack, swipe a card to pick up a vehicle, and drive. They then drop off the vehicle at the stack closest to their destination.

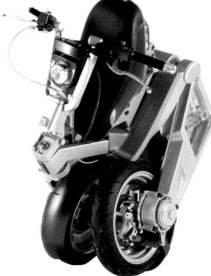

© Smart Cities Group/MIT Media Lab (all images p118-119)

CASE STUDY: FLORENCE, ITALY

Large vehicle storage areas can lie outside the dense historic center.

Major mobility nodes exist at the traditional city gates.

Minor mobility nodes are aligned with piazzas and existing transportation hubs.

Minor "snap-on" stacks and racks can be placed in streets and adjusted over time.

0 100   250   500   1000m

© Smart Cities Group/MIT Media Lab

GREEN CORRIDOR

Green corridors in a city's center offer a place of respite for residents, commuters and visitors. Recently, New York City opened the High Line, a very successful urban park built above grade on an abandoned freight rail line built in the 1930s.

The High Line incorporates a 1.5-mile-long elevated park that runs through the West Side neighborhoods of the Meatpacking District, West Chelsea and Clinton/Hell's Kitchen. It features a landscape combining meandering concrete pathways with local, low-maintenance plantings. Fixed and movable seating, lighting and special features are also included in the park. Access points from the street level are located every two to three blocks. Many of these access points include elevators.

In the southern area of Madrid, part of the Park Manzanares is a snaking path called Paseo de los Sentidos, which is planted with palm, oak, cork and olive trees, along with and other Mediterranean species. It incorporates an amphitheatre, public art and places to relax. The full path is about seven kilometers long, and follows the course of the river while still being incorporated into the urban fabric.

Funding

There is the potential to explore the use of the existing provisions of the Illinois TIF (Tax Incremental Financing) law that permit the use of tax increment financing for inter-modal transport.

Ideally, this could involve a large inter-modal district encompassing our study area, McCormick Place and the Museum Campus to the south, the University of Illinois and the medical district to the west and River North, Michigan Avenue, Streeterville and Navy Pier to the north.

In addition to such an inter-modal district, consideration can be given to supplementing revenue with a transit-based business improvement district, with a number of federal programs designed to encourage mobility and multiple modes of transport.

1 Las Ramblas, Barcelona
2 Park Manzarares, Madrid
3 The High Line, New York

MONROE EAST–WEST INTERMODAL AXIS

In response to the analysis of the existing transportation options and proposed items in the Chicago Central Area Action Plan, a series of design strategies evolved as a means to integrate the existing and the new into a coherent, functional and thriving urban system.

We propose that a multi-level East–West intermodal axis be created along Monroe Street. This idea is consistent with current master planning for the central Loop. The Monroe corridor has the potential to be a convenient link for transit and pedestrians, and a thriving retail space. Its central location in the Loop will make it a hub for commuters.

A below-grade electric bus is proposed that turns around at Millennium Park, connecting Ogilvie and Union Stations on the west, and Blue and Red Lines, with future connections to O'Hare, Grant Park and the Millennium Station trains. There is no below-grade infrastructure along Monroe, although it has been envisioned as a future transportation connection since the 1960s. At present, Monroe is three lanes wide with no parking or stopping lanes. In our proposal, Monroe is narrowed to two lanes, creating a pathway above grade that would incorporate pedestrian boulevards, bicycle paths and vegetated recreational areas. There is also potential to eventually create a connection to the future Kennedy Green Cap Plan.

There are currently breaks in the Loop pedway system that could be connected to make one continuous below-grade system. Below grade, the Monroe East–West intermodal axis provides a covered space during inclement weather. This space can be programmed to incorporate amenities and

services such as shopping, gyms, entertainment, bicycle stations, a library, spas, grocery stores and restaurants. Considering the extreme Chicago weather conditions, there is potential for building owners to capitalize on their connection to a thriving pedway system. This will encourage potential retail and services to cater to increasing Loop populations. A connection to the proposed East–West Monroe link could aid in optimizing this pedway. There are locations, such as Chase Plaza, where the below-grade portion erodes and connects with the above-grade streetscape. There are also interconnections up and down to the shuttle bus stops.

The bicycle/pedestrian paths along the river's edge can be continued on both sides of the river. There are areas where existing wider spaces can incorporate further amenities such as performance spaces and cafes. The path can also connect to the Monroe East–West intermodal axis to accommodate commuters biking the Last Mile/First Mile portions of their commute.

The green axis is the main spine that links a series of smaller green plazas that support Loop inhabitants during the day and the evening. The green connections improve transportation, amenities and livability; reduce the heat island effect and storm water treatment requirements; and increase air quality.

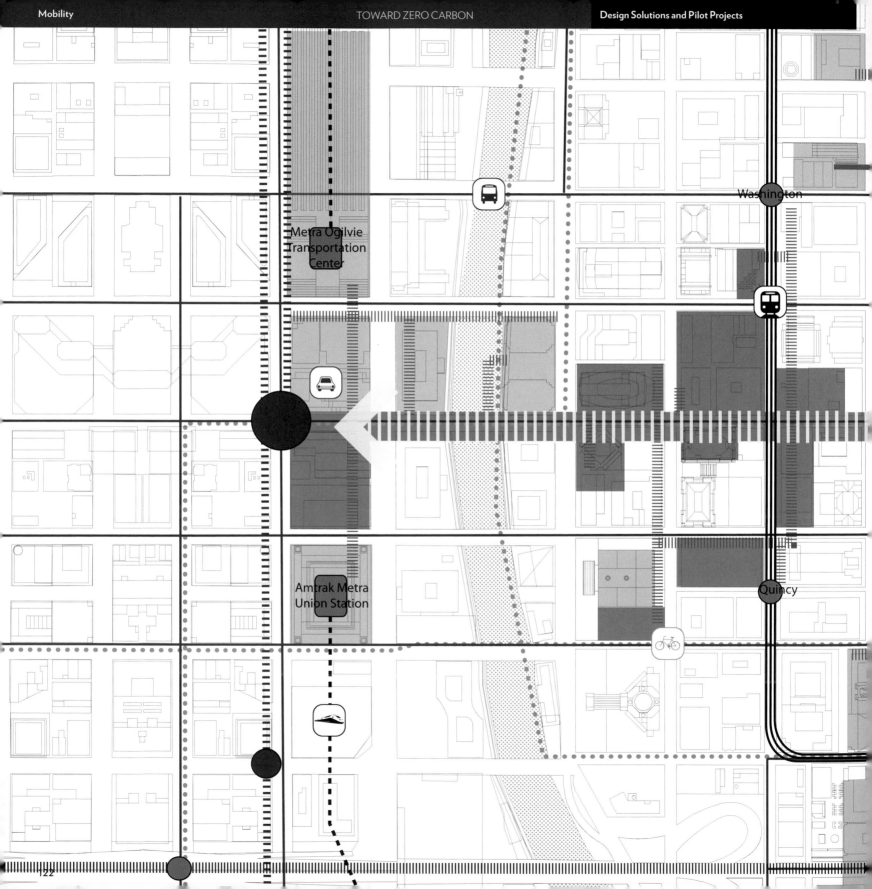

Washington

Metra Ogilvie
Transportation
Center

Amtrak Metra
Union Station

Quincy

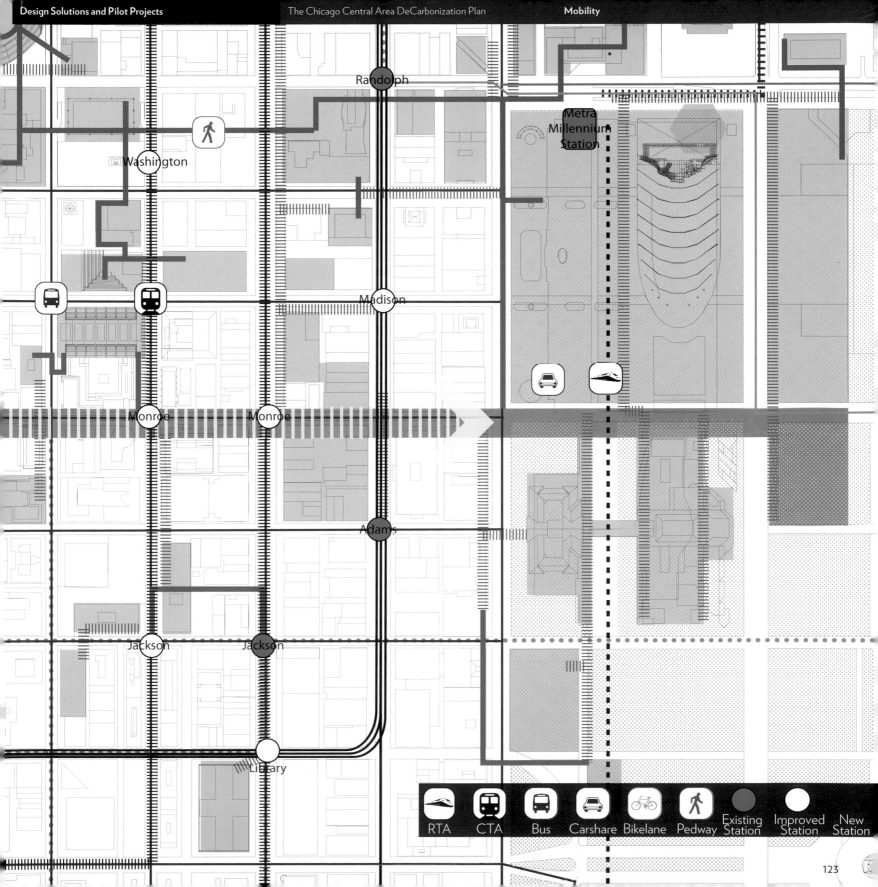

Randolph

Metra
Millennium
Station

Washington

Madison

Monroe   Monroe

Adams

Jackson   Jackson

Library

RTA   CTA   Bus   Carshare   Bikelane   Pedway   Existing Station   Improved Station   New Station

INTERMODAL GREEN AXIS

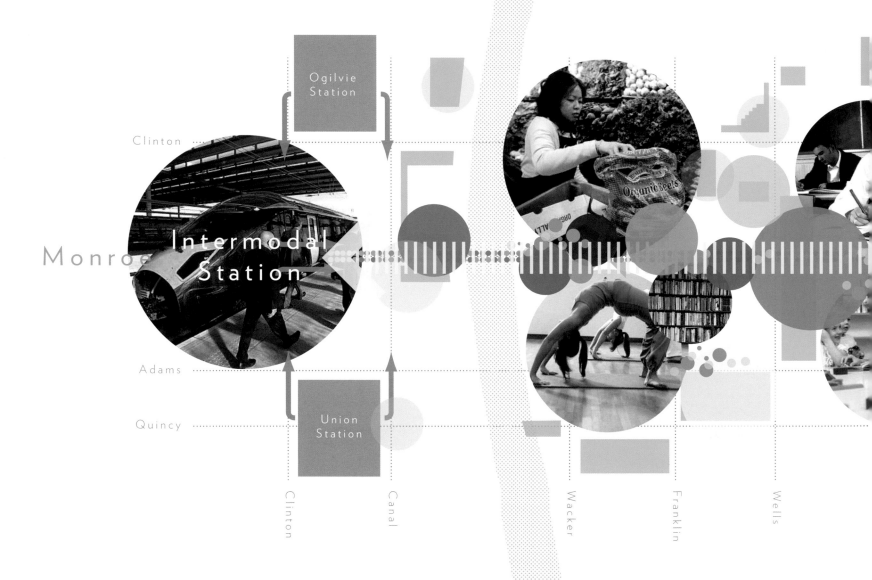

Clinton

Ogilvie
Station

Monroe

Intermodal
Station

Adams

Quincy

Union
Station

Clinton

Canal

Wacker

Franklin

Wells

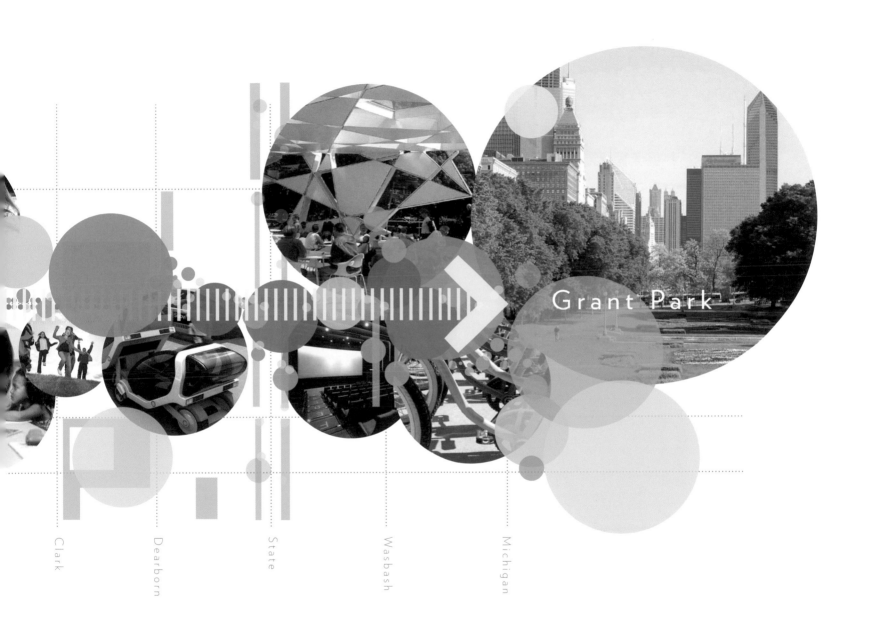

Grant Park

Clark

Dearborn

State

Wasbash

Michigan

INTERMODAL GREEN AXIS

INTERMODAL AXIS ABOVE GRADE

Intermodal electric bus line below grade
with connections to amenities and pedway.

PEDESTRIAN AND BICYCLE AMENITIES

The bicycle/pedestrian paths along the Chicago River's edge can be continued on both sides of the river. There are areas where existing wider spaces can incorporate further amenities such as performance spaces and cafes. The bicycle path can also connect to the Monroe East–West Intermodal Axis to accommodate commuters biking the Last Mile/First Mile portions of their commute.

The Loop currently has one bicycle station on the east side of Randolph. This station does supply services such as storage and repairs to commuters, but its primary focus is rentals to tourists. A new, larger facility closer to commuter connection would encourage bicycle commuting.

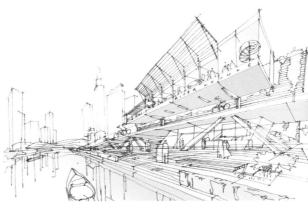

135

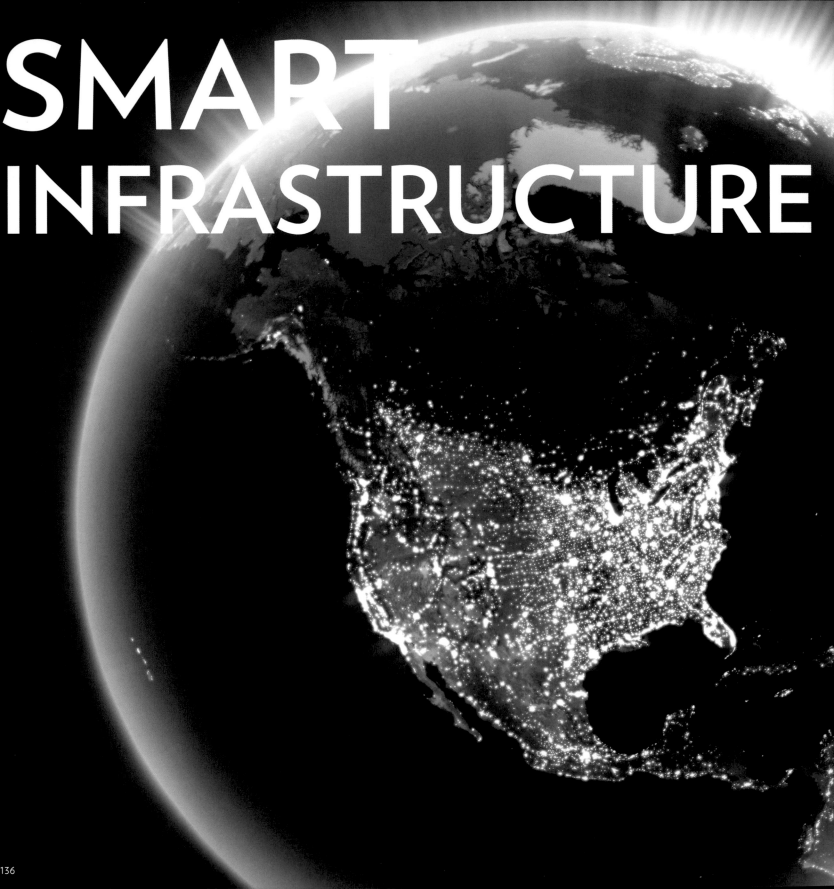

SMART
INFRASTRUCTURE

An integral component of a modern clean-energy paradigm will be an intelligent supply/demand/ storage network that increases reliability and minimizes carbon emissions. As Illinois and Chicago increasingly rely on intermittent renewable energy sources, this approach will go from a value-add to a necessity.

SMART INFRASTRUCTURE
Carbon Reduction Strategies

What if more than our energy supply could be intelligent? What if the whole social infrastructure of our cities could be enabled through information technology? Energy consumption in the United States, and correspondingly in the city of Chicago, has dramatically escalated over the past several decades, significantly outpacing population growth as new energy consumers such as televisions, computers, mobile phones and iPods have become ubiquitous in our daily lives. While these technologies have led to increased strain on an aging infrastructure, they also present new opportunities to improve the intelligence and distribution efficiency of energy and information, engendering new infrastructure intelligence.

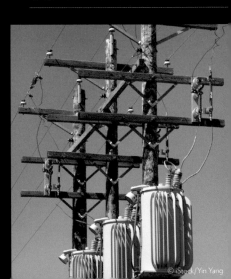

In **Analysis**, the regional trends in energy supply for the state of Illinois and the city are studied in aggregate and broken down by energy source—coal, natural gas, nuclear, hydroelectric, wind and other renewable sources. These renewable sources are projected to contribute an increasing percentage of the city's energy supply in the years to come. Along with this increased reliance on low-carbon technologies will come additional demands on our energy distribution infrastructure, requiring deployment of smart-grid technologies.

In the **Strategies** section, we expand the concept of the smart grid to encompass the broader scope of smart infrastructure, which goes beyond supply-and-demand management strategies and seeks to add amenities to the city through information technologies.

In the **Design Solutions and Pilot Projects** section, we consider five distinct networks: real estate, retail, public realms, transit and utilities. Within these networks, we identify specific Chicago projects, such as a clean energy taxi program and a vehicle-to-grid car-sharing network.

© iStock/Yin Yang

CARBON REDUCTION GOALS

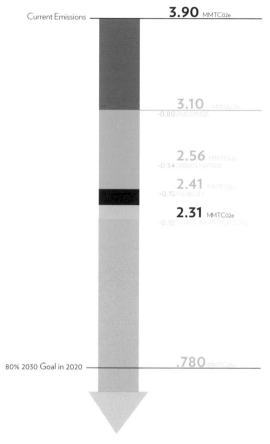

Current Emissions — **3.90** MMTC02e

3.10 MMTC02e
-0.80 BUILDINGS

2.56 MMTC02e
-0.54 URBAN MATRIX

2.41 MMTC02e
-0.15 MOBILITY

2.31 MMTC02e
-0.10 SMART INFRASTRUCTURE

80% 2030 Goal in 2020 — **.780** MMTC02e

There is no specific carbon quantity related to infrastructure. It supports the goals of the other elements of the plan.

The U.S. electricity grid is a remarkable infrastructure worth over $1 trillion. It includes more than 200,000 miles of high-voltage transmission lines and 5.5 million miles of distribution lines servicing hundreds of millions of end users. As it is uneconomical with current technology to store a significant amount of energy, the grid must be constantly monitored and balanced in real time throughout the day. The resulting inefficiencies lead to increased carbon emissions, as power plants run on part load conditions or lose significant amounts of energy in distribution. Fortunately, these characteristics can be minimized with intelligent control systems.

Intelligent infrastructure can also have broader implications with synergies in a number of areas such as:
· Continuous commissioning of buildings
· Reduced transmission and distribution (T&D) line losses
· Direct feedback to customers
· More effective and reliable demand response and load control
· Enhanced measurement and verification (M&V) capabilities

Average cost for 1 hour of power interruption:

Cellular communication	$41,000
Telephone ticket sales	$72,000
Airline reservation system	$90,000
Credit card operation	$2,580,000
Brokerage operation	$6,480,000

Today's electricity system is 99.97% reliable, yet still allows for power outages and interruptions that cost Americans at least $150 billion each year—about $500 for every man, woman and child.

In 1988-98, U.S. electricity demand rose by nearly 30%, while the transmission network's capacity grew by only 15%. Summer peak demand is expected to increase by almost 20% during the next 10 years.

REGIONAL ENERGY TRENDS

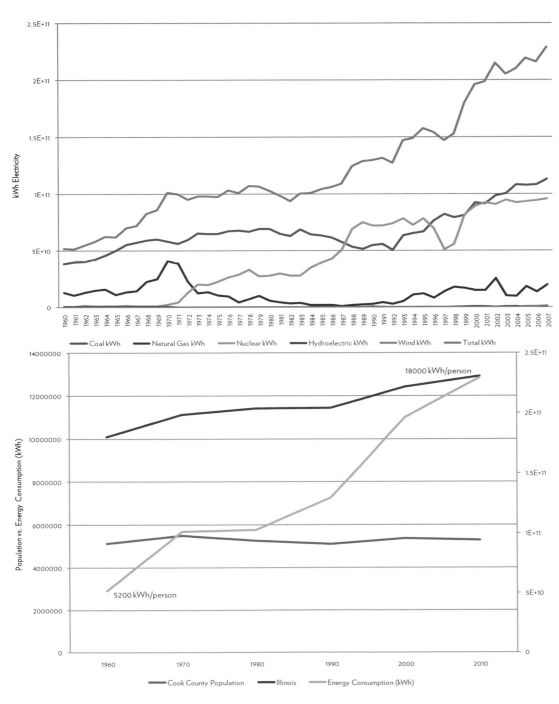

Energy production (and therefore carbon emissions) has accelerated over the past half-century, corresponding to increased demands due to the widespread use of the personal electronics that typify our information age. Taking into account pending federal legislation to curb greenhouse gas emissions and increase the energy efficiency of systems, we will soon rely more heavily on low-carbon energy technologies. These include traditional hydrocarbon-based sources with carbon capture and sequestration methodologies, combined heat and power (CHP) applications of hydrocarbon sources and renewable energies such as wind and solar. The reliance on these technologies will necessitate new means of managing energy supply and demand.

ENERGY COSTS AND FUTURE PROJECTIONS

On an inflation-adjusted basis over the last 50 years, energy costs in the state of Illinois are currently some of the lowest, making the valuation of carbon emissions necessary to justify the economics of many systems discussed in other sections of this book. A number of recent studies have concluded that simple behavioral psychology techniques, such as informing users of their energy consumption relative to their peers, can lead to 10-25% reductions in demand. A smart infrastructure network can facilitate such strategies.

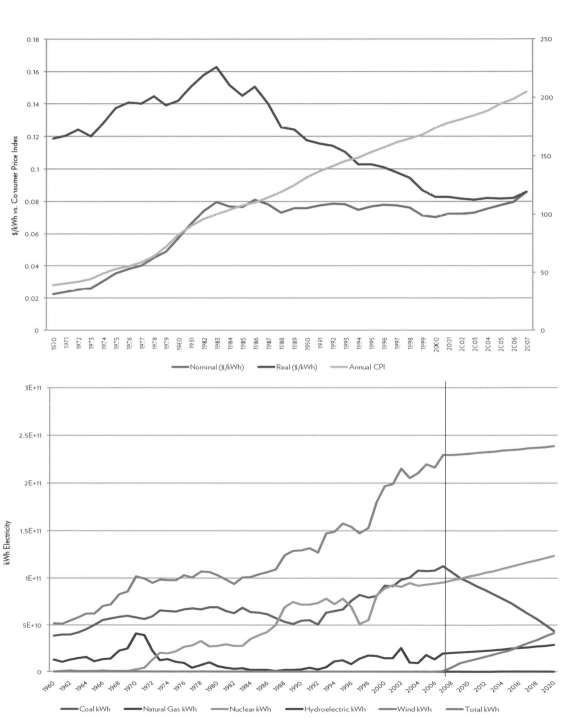

SMART INFRASTRUCTURE OVERVIEW

UBIQUITOUS SYSTEM

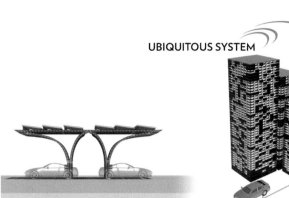

"SMART-GRID" COMMUNICATION NETWORK

- Photovoltaic sources
- Heating/Cooling load sources
- Wind power sources

ENERGY STATION

PHEV plug-in charging system

SMART CHARGING

intelligent management of energy from source to vehicle

CENTRALIZED SYSTEM

A data center / thin-client solution can improve flexibility and reduce power demands by over 200 W/workstation.

Funding

The Funding chapter describes a number of programs that encourage smart infrastructure investment within new and existing buildings, and also in the realm of classic infrastructure such as roads, transit, sewers and water.

An array of grants and tax credits, principally at the federal level, exists to help foster the adoption of smart infrastructure. Please see the appendix to the Funding chapter for descriptions of a number of these incentives. Moreover, several programs exist in the U.S. Department of Housing and Urban Development (HUD), the U.S. Environmental Protection Agency and the U.S. Department of Transportation.

Under the direction of the Livable Communities Office, the Obama Administration has charged the secretaries and administrator of these agencies to work together, making the sum of their efforts greater than their individual parts.

Smart infrastructure provides one of the best ways to achieve energy savings because it touches many parts of an urban context, including buildings and public improvements and their interconnectedness.

High-speed connections
Advanced sensors distributed throughout the grid and a high-speed communication network tie the entire system together.

Energy storage and smart demand system
Smart-grid systems integrate hardware, software and services to intelligently manage power demand. For demand response applications, smart-grid systems allow for utilities to remove the power to certain appliances during peak times so that the utility can produce power more efficiently. These systems can also enable solar power generation by dynamically switching supply from solar to micro-turbines or fuel cells.

Plug-in hybrid electric cars
Plug-in electric vehicles can store energy and, as the majority of the Loop buildings are offices, peak demand may be offset by using plug-in hybrid electric vehicles.

Smart appliances
Smart appliances contain on-board intelligence that "talks" to the grid, sensing grid conditions and automatically turning devices on and off as required.

Smart thermostats
Customers can opt to use a smart thermostat, which can communicate with the grid and adjust device settings to help optimize load management. Other smart devices could control air conditioners and pool pumps.

Customer choice
Customers could be offered an opportunity to choose the type and amount of energy they would like to receive with the click of a mouse on their computers.

SMART GRID APPLICATIONS

© iStock/Stephen Strathdee

Intelligent Demand Management

For the utility
- Reduce peak demand through a two-way wireless thermostat
- Events can be scheduled in advance or conducted in real time (less than 5 minutes)
- Verification of demand reduction and participation
- Leverage network operation center statistics, master data management integration and back-office systems

For the customer
- View general energy use and costs
- Receive pending notification of demand response events with opt-out option

Load Measurement and Control

For the utility
- Measure power consumption by circuit
- Predict, control and verify load curtailment
- Events can be scheduled in advance or conducted in real time (less than 5 minutes)
- Typical sustainable load curtailment of 1 kW-3 kW per household

For the customer
- Savings: 10-15% reduction on household electric bill
- Visibility: detailed circuit-level consumption data
- Control: personal energy profile

Energy Storage Per Household

For the utility
- Stored energy at point of consumption can be discharged to provide peak power
- Currently 10 kWh of stored energy with a 3.3 kW maximum discharge rate
- Events can be scheduled in advance or conducted in real time (less than 5 minutes)

For the customer
- Instant, clean, silent backup power with remote monitoring
- Protect critical and sensitive loads such as computers, refrigerators and heating systems
- Optional generator interface provides long-term outage protection
- For light commercial applications, stored energy

SMART GRID ELEMENTS

Smart-grid systems integrate hardware, software and services to intelligently manage power demand. They allow users to communicate on-the-fly information about the current price and use of electricity.

1. **Solar panels** mounted on buildings generate power during the day. If a building is generating surplus power, it can be fed back to the utility company and be reimbursed.

2. **Wind turbines** spin and generate power from air movement. Like solar panels, wind turbines also have the potential to generate a power surplus.

3. **Smart appliances** monitor how much electricity they're using and shut down when power is too expensive.

4. **Remote control** consumers can permit utilities to control their non-essential appliances such as pool pumps, turning them on and off to fine-tune the grid for maximum efficiency.

5. **Plug-in hybrid cars** refuel using clean electricity generated locally.

6. **Locally generated power** avoids the 15% power loss that occurs when electricity is sent over long-distance power lines. "Superconducting" power lines route extra electricity from out-of-state utilities when demand spikes.

7. **Wireless chips** let individual houses communicate with utilities, swapping on-the-fly information about the current price and use of electricity.

8. **Web and mobile-phone interfaces** allow consumers to see how much power their appliances are using when they're not at home, and even to turn them on or off remotely to reduce costs.

9. **Energy storage.** When solar panels produce excess energy, it can be stored in batteries so that houses can use clean energy at night when the sun isn't shining.

FORWARD-THINKING CITIES

Beach Cities MicroGrid
by San Diego Gas & Electric

As its name implies, a micro-grid resembles our current grid, although on a much smaller scale. It has the unique ability, during a major grid disturbance, to isolate from the utility seamlessly with little or no disruption to the loads within it and reconnect later, also seamlessly.

The Beach Cities MicroGrid Project will be demonstrated at an existing substation identified as "Beach City Substation." It's intended to offer a blueprint to all distribution utilities, proving the effectiveness of integrating multiple distributed energy resources with advanced controls and communications. It seeks to improve reliability and reduce peak loads on grid components such as distribution feeders and substations. Both utility-owned and customer-owned generation (such as photovoltaic systems and biodiesel-fueled generators) and energy storage will be integrated, along with advanced metering infrastructure (AMI), into the real-world substation operations with a peak load of approximately 50 MW.

Beach Cities will serve as a guide for improved asset use, as well as for operating the entire distribution network in the future. Successfully building such capabilities will enable customer participation in reliability- and price-driven load management practices, both of which are key to realizing a smarter grid.

High penetration of clean energy technologies
by the city of Fort Collins

The city and its city-owned Fort Collins Utility support a wide variety of clean energy initiatives, including the establishment of a Zero Energy District within the city (known as FortZED). One such initiative seeks to modernize and transform the electrical distribution system by developing and demonstrating an integrated system that distributes mixed resources to increase the penetration of renewables—such as solar and wind—while delivering improved efficiency and reliability.

These and other distributed resources will be fully integrated into the electrical distribution system to support the Zero Energy District. In fact, this U.S. Department of Energy-supported project involves the integration of a mix of nearly 30 distributed generation, renewable energy, and demand response resources across five customer locations for an aggregated capacity of more than 3.5 MW.

This project will help determine the maximum degree of penetration of distributed resources based on system performance and economics.

Perfect Power
by Illinois Institute of Technology (IIT)

A "Perfect Power" system is defined as an electric system that cannot fail to meet the electric needs of the individual end-user. A Perfect Power system has the flexibility to supply the power required by various types of end-users and their needs without fail. The functionalities of such a system will be enabled by the smart grid.

This project will design a Perfect Power prototype that leverages advanced technology to create micro-grids that respond to grid conditions and provide increased reliability and demand reduction. This prototype model will be demonstrated at the IIT campus to showcase its operations to the industry. The model is designed to be replicable in any municipality-sized system where customers can participate in electric market opportunities.

SMART REAL ESTATE

Every building has unique challenges and opportunities. Therefore it makes sense to address and optimize buildings on an individual basis prior to integration with a smart grid network—the "smart nodes on a smart grid" concept. Information technology can be used beyond traditional building management systems to provide services to enhance tenant experience, such as high-speed communication and data management, carbon-emission accounting and other potential performance objectives of corporations today.

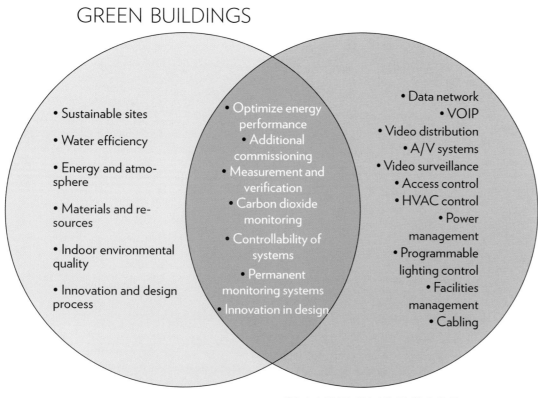

GREEN BUILDINGS

- Sustainable sites
- Water efficiency
- Energy and atmosphere
- Materials and resources
- Indoor environmental quality
- Innovation and design process

- Optimize energy performance
- Additional commissioning
- Measurement and verification
- Carbon dioxide monitoring
- Controllability of systems
- Permanent monitoring systems
- Innovation in design

- Data network
- VOIP
- Video distribution
- A/V systems
- Video surveillance
- Access control
- HVAC control
- Power management
- Programmable lighting control
- Facilities management
- Cabling

SMART BUILDINGS

A data center/thin client solution can improve flexibility and reduce power demands by over 200 watts per workstation. This system is purpose-built for server-based desktop virtualization. It moves all software off the desktop to the server, so the device has no CPU, no persistent memory, no operating system and no drivers. Using functionality enabled by server virtualization, this solution delivers not only a complete experience, including native USB support, but also a new set of high-value features not available with current desktop and thin client architectures.

Technologies

· By moving all software off the desktop and onto the server, this system reduces desktop TCO by 70%, saving as much as $3,200 per desktop over three years.
· Physical visits to the desktop endpoint by IT personnel are eliminated.
· Software installations and upgrades are performed centrally.
· These devices consume less than 5 W—3% of the power used by a typical PC—making them the ultimate green alternative.

Environmental controls and next generation facilities solutions

Thanks in part to Internet protocol-based networks, new digital technologies are ready to make dramatic contributions to how buildings function, particularly in reducing their energy consumption. Though it might still sound a bit far-fetched for bricks and mortar to have a brain, industry experts say today's technology is now more than capable of giving buildings this kind of intelligence.

A smart building can be almost any structure, from a shopping mall or home to a hospital or an office high-rise. They all share the common ability of "knowing" what's going on inside their walls and being able to respond accordingly. Smart buildings control the building automation systems for monitoring

and regulating heating, air conditioning, lighting and other environmental variables. They can also oversee other building functions such as security, fire suppression and elevator operations.

Beyond integration, smart building technologies focus on bringing more detailed monitoring and sensing "awareness" to buildings. Typically, heating and cooling systems have one thermostat for an entire office floor. But new smart building networks can now cost-effectively provide far more detailed monitoring of the conditions inside a building, helping a structure's environmental systems deliver just enough heat, air or cooling when and where it's needed. Smart buildings equipped with an integrated array of sensors can also monitor such things as the amount of sunlight coming into a room and adjust indoor lighting accordingly. Advanced smart buildings can know who is visiting a building after hours (based on key swipes from the security system) and turn on the appropriate lights, equipment and environmental controls.

The bottom-line result of this intelligence and coordination is much lower operational costs for commercial buildings. Industry research suggests that such overheads can account for as much as 80% of the cost of a building during its lifetime, including construction. Much of that expense is from energy use.

Power over ethernet thin client solution
– Flexible and secure
– Ultra-low energy/cooling

Bespoke environmental controls
– Natural light spectrum LEDs
– Demand response ventilation
– Self-powered sensors and controls

Intelligent facilities management
– Continuous commissioning
– Carbon management ISO 14001 ready

SMART RETAIL

Information technology presents a unique opportunity for retailers to move beyond the tried-and-true focus areas of the past: location, product and supply-chain efficiency. The new frontier of opportunity will enable concepts such as individual profitability, lifetime value, customer loyalty and personalization. In this way, the objectives of sustainable development—environmental, social and economic—are supported.

Effective application of these concepts significantly challenges many traditional retail models. Today's growth opportunities require new organizations, processes, technologies and, most importantly, new mindsets.

Technology-enabled efficiency frees resources to create high-value interactions

- Self-checkout
- Digital signage
- Web-based services and channel integration
- Self-service kiosks
- Product sensing
- Electronic shelf labeling
- Facilitate interactions through knowledge-based selling
- Deploy technology to provide expertise about products and individual customers at point of purchase
- Rich media e-learning
- Handheld devices with customer and product information for sales associates
- Expert kiosk—live and virtual product experts "click to chat"
- Customer sensing—Internet and physical presence

SMART PUBLIC REALMS

A smart public realm is a social infrastructure service made more intelligent by information technology and broadband connectivity. Potential services that could enhance the public experience of the city range from parks with wi-fi access, information kiosks for tourists and interactive media outlets to billboards, security and emergency response systems. These services are critical to attracting and retaining inhabitants, thus promoting the economic vitality and sustainability of the city.

City lifestyle amenity center—establish carbon-neutral social infrastructure (recreation, education, food)

Public/private partnership development
- Revenue from food market
- Revenue from year-round recreational facility
- Possible federal support for preventative health care

Recreation
- Facilities for students
- Facilities for members

Food market
- Facilities for public
- Urban garden
- Online ordering and private nutrition monitoring

Education
- Potential charter school
- General K-12 or targeted

Smart Parks—maximum educational impact with minimal environmental loadings

Park Version 1.0
- Amenity = lawn
- Energy = near zero

Park Version 2.0
- 600 gpm fountains
- 50 foot LED displays
- 113- to 1200-watt loudspeakers

Park Version 3.0
- Interactive, personalized content delivery
- Integrated renewable energy
- Enhanced carbon sequestration

SMART TRANSIT

Transportation is Chicago's second-largest contributor to greenhouse gas emissions, and as such presents a major opportunity to reduce carbon and improve the overall logistics of travel through information technology services. In the future, vehicles in tandem with a plug-in hybrid system could present opportunities for energy storage. Beyond that, intelligent infrastructure could help navigate traffic, find a destination and increase safety during travel.

PHEV CHARGE PLUG   PHEV USER INTERFACE

SMART METER

BUILDING INFRASTRUCTURE   ELECTRIC GRID   RENEWABLE ENERGY SOURCES

Clean government vehicle procurement
- Electric/natural gas fleet vehicles
- Clean cities program
- Eliminate garbage trucks from Loop though pneumatic conveyance

Establish policy that promotes private electric vehicle ownership
- Payment models for regulation and reservation services
- Parking/congestion charging
- Space locator for trolling prevention

Taxi fleet emission standards
- Establish market for clean cab services
- Enable premium services for tourism and marketing

© iStock/Stephan Zabel

SMART GRID UTILITIES

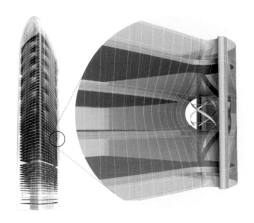

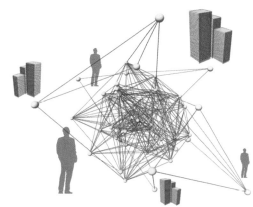

Renewables integration

For the utility
- Utility control over distributed renewable energy sources
- Measure and verify production to predict supply, accumulate RECs and meet ROS mandates
- Ability to stabilize intermittency by automatically dispatching distributed stored energy
- Sell or lease renewable energy systems to customers or co-market with third-party providers

For the customer
- Purchase utility-grade renewable energy system from a trusted provider—their power company
- 40-60% savings on electricity bill
- Grid point customer portal provides comprehensive production and environmental impact data

Plug-in hybrid vehicles

For the utility
- New source of revenue (smart-charging)
- Optimize base load generation
- Value-based pricing to control when and how fast PHEVs will recharge
- Stored energy in PHEVs can be discharged to provide peak power (future)
- Measure and verify discharged power for billing reconciliation with customers

For the customer
- Reduce cost of electricity used by PHEV
- Utilize PHEV to participate in demand response programs
- Grid point customer portal provides comprehensive production and environmental impact data

Distributed generation

For the utility
- Remotely dispatch power from commercial generators into the grid during demand response events
- Events can be scheduled or on-demand
- Measure and verify dispatched power for billing reconciliation with customers

For the customer
- Utilize standby generators to participate in demand response programs in combination with load curtailment activity
- Utilize standby generators to mitigate peak demand charges

WATER

The carbon dioxide embedded in the nation's water represents 5% of all U.S. carbon emissions and is equivalent to the emissions of more than 62 coal-fired power plants. Even a city situated on the shores of one of the world's largest sources of fresh water—Lake Michigan—needs to find ways to reduce water consumption and the carbon load per gallon.

It's difficult for many Chicagoans to imagine that the high-quality water they enjoy might be a source of greenhouse gas emissions. To a degree, this mindset is understandable, given the proximity and relative purity of our main water source, Lake Michigan. But the purification, delivery, heating and treatment of waste water can be significantly improved to reduce the overall carbon footprint of water in Chicago.

In **Analysis**, we examine water use nationally, to establish a larger context, then look at use on the city level. A water cycle is discussed with the associated carbon load of each step in the cycle, framing the strategies for carbon reduction in water.

In the **Strategies** section, two general areas are studied to begin to investigate discrete methods for carbon reduction. We look at reducing the carbon load in a gallon of water, then examine conservation efforts to create a cascading strategy for carbon reduction.

Design Solutions and Pilot Projects presents several possibilites for the carbon loading and conservation sides of the carbon reduction effort for water. Solutions for the use and conservation of

this valuable natural resource range in scale from the replacement of aerators on sinks to the restoration of wetlands.

CARBON REDUCTION GOALS

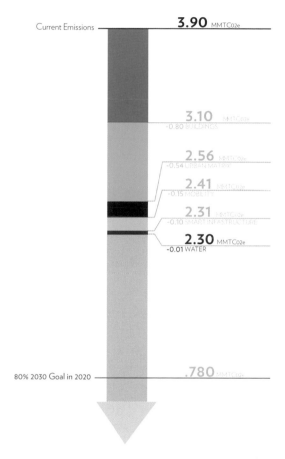

Current Emissions — **3.90** MMTC02e

3.10 MMTC02e
-0.80 BUILDINGS

2.56 MMTC02e
-0.54 URBAN MATRIX

2.41 MMTC02e
-0.15 MOBILITY

2.31 MMTC02e
-0.10 SMART INFRASTRUCTURE

2.30 MMTC02e
-0.01 WATER

80% 2030 Goal in 2020 — **.780** MMTC02e

Specific goals as related to water emission reduction within the DeCarbonization Plan

Many other major metropolitan areas don't enjoy the luxury of a seemingly infinite water source at their disposal. Almost half of the water used by the city of Los Angeles, for example, is pumped from more than 700 miles away, over the 2,000-foot Tehachapi Mountains. The other half flows by gravity from the Sierra Nevada mountains, but the aqueduct that had to be constructed for this is more than 400 miles long. About 40% of New York City's water supply flows through the Catskill Aqueduct, which is 163 miles long and varies in depth from 174 to 1,187 feet. Even though Chicago's delivery system may seem simple in comparison to these other major metropolitan examples, the sheer volume of water delivered and treated—almost as much as in New York City—creates a significant source of carbon that has many opportunities for reduction.

Determining water use intensity in the study area

We derived study area water use intensity using data collected from pilot buildings. We then compared use intensity to total city pumpage through floor area ratios to determine the water use of the study area.

- Study area water use intensity—13.03 gal/sf/yr
- Study area water use—1,563,201,726 gal/yr (.481% of total city pumpage)

Determining conservation required to meet the DeCarbonization Plan goal

Initial water-borne carbon reduction goals were derived from the Chicago Climate Action Plan. CCAP mitigation Action 4, "Conserve Water," has a .04 MMTCO2e reduction target for the entire city. To determine carbon load, the Chicago Department of Water Management provided estimates of power associated with purification, delivery, end-user heating and treatment, which were converted to carbon to arrive at a carbon-per-gallon approximation for all Chicago pumpage. This carbon load was then divided into the pro-rated CCAP carbon reduction goal associated with water conservation in the study area to arrive at a total gallon requirement to meet the goal:

- Chicago carbon load/gal—.005133 lb/CO2e (DWM estimate)
- Carbon goal for study area—424,348 lb/CO2e
- Required conservation to meet goal—82,670,523 gal/yr (5.3% reduction in use intensity)

To clarify, this amount of water conservation, based on the current estimated carbon load per gallon, is sufficient to meet the CCAP goal for the study area. Other strategies already proposed or underway will further reduce the carbon load per gallon to increase the carbon savings realized through conservation.

CCAP REDUCTION GOAL
0.04 MMTCO2e

CLD REDUCTION GOAL
.01 MMTCO2e   (30%)

CHICAGO 2006 WATER USE
325 billion gallons

ESTIMATED STUDY AREA 2006 USE
1.6 billion gallons

Less than 1% of the world's fresh water is available for human consumption.

NATIONAL WATER USE

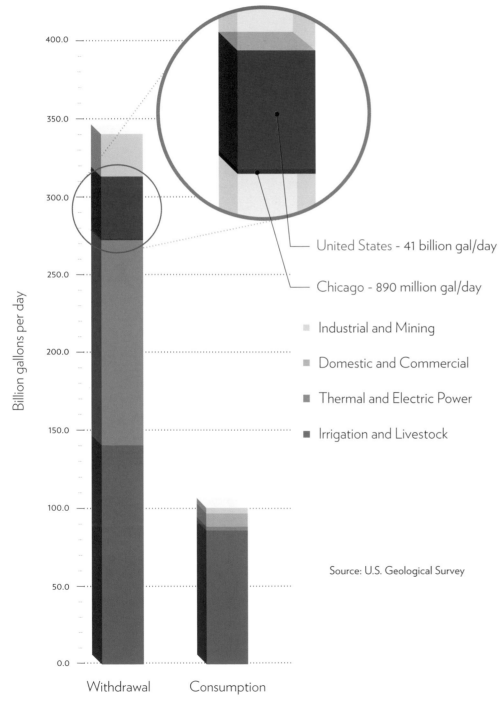

United States - 41 billion gal/day

Chicago - 890 million gal/day

■ Industrial and Mining

■ Domestic and Commercial

■ Thermal and Electric Power

■ Irrigation and Livestock

Source: U.S. Geological Survey

Billion gallons per day

Withdrawal   Consumption

Although more water is withdrawn for thermal and electric power than any other use, 97% simply passes through heat exchangers and is returned to the source. Conversely, domestic and commercial withdrawals go through several carbon-loading processes before being delivered for use. Even more carbon-intensive processes are involved on its return, in the form of waste water treatment.

CHICAGO WATER USE AND CARBON LOAD

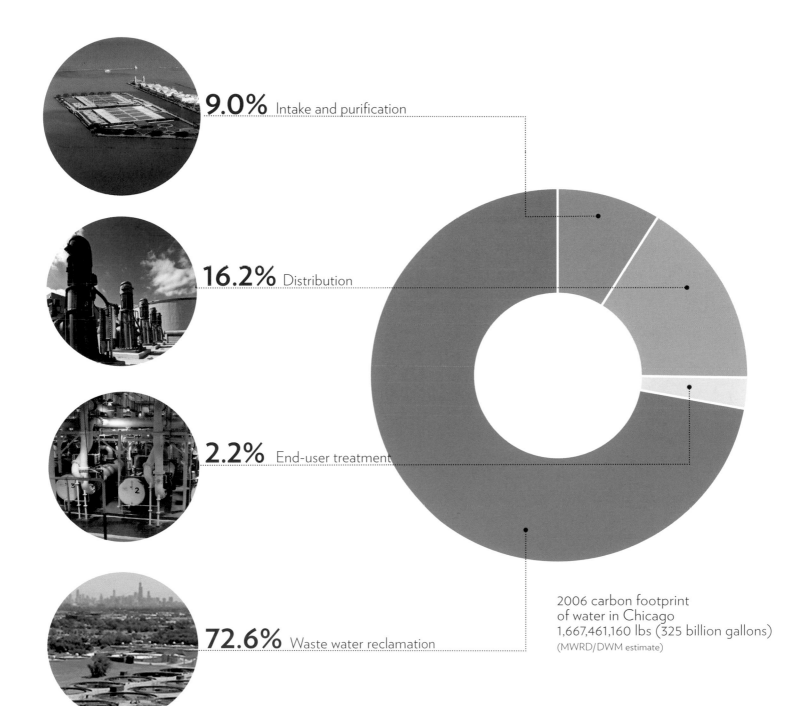

9.0% Intake and purification

16.2% Distribution

2.2% End-user treatment

72.6% Waste water reclamation

2006 carbon footprint
of water in Chicago
1,667,461,160 lbs (325 billion gallons)
(MWRD/DWM estimate)

CARBON REDUCTION METHODS IN WATER

The strategies for carbon reduction from water fall into two basic categories: (1) reducing carbon load per gallon and (2) water conservation.

Waste water treatment represents 72.6% of the carbon load of a gallon of domestic water in Chicago.

Reducing carbon load per gallon

The Chicago Department of Water Management (DWM) is aggressively pursuing increased efficiency in its purification and water delivery systems. Old and inefficient steam-driven pumps are being converted to high-efficiency electric motors with variable speed drives. Existing electric motors are also being upgraded to the same standard. On the waste water treatment side, the Metropolitan Water Reclamation District is upgrading motors to the same high efficiency standards as DWM. The department is also investigating the installation of renewable energy sources at pumping stations through a pilot project at its Lexington Station and is considering renewable wind energy at intake cribs on Lake Michigan.

End users' contribution to the carbon load of water can also be improved. Many commercial office buildings are placing their large boilers on time clocks so that hot water is only heated and recirculated during obligated lease hours. Others are taking matters further and replacing large centralized boiler systems with smaller local water heaters. This strategy allows more point-of-use supply, which eliminates the need to recirculate hot water through long risers to ensure that hot water is always available at the faucet. If small enough,

as with undercounter water tanks used for office pantry kitchens, the electrical demand can be supplied from the tenants meter, making the tenant more aware of the cost of use and often resulting in higher levels of conservation.

Waste water treatment is the real culprit in the carbon load of water, accounting for nearly three-fourths of the carbon load of a gallon of water in Chicago. Although retrofitting of pumps and motors at treatment plants is a good start, alternative treatment methods are needed to significantly reduce the carbon contribution of waste water treatment. These alternative methods of treatment range in scale from small, on-site "living machine" filters to wetland restorations covering tens of thousand of acres. These biological systems often completely supersede the need for brick-and-mortar plants, and provide a range of other benefits such as increased biodiversity. Smaller, less energy-intensive mechanical systems such as membrane bioreactors are starting to reach a level of development that will allow them to be used on a municipal scale.

Conservation

After reducing the carbon load of each gallon of water, conservation creates a cascading effect of carbon reduction.

Replacement of all toilets, urinals and faucets in the study area to post-1994 fixtures would result in an annual saving of 861 million gallons/yr, over ten times the conservation goal.

CONSERVATION

At the municipal scale

At a large scale, the Chicago Department of Water Management has implemented an aggressive program to reduce leakage in distribution piping. Distribution losses are currently estimated at 10%. Reduction by half of this amount would result in a carbon saving of .76 MMTCO2e per year. The DWM program seeks to inspect every water main every three years and increase its water main replacement target to 75 miles per year by 2015.

Metering may be commonplace everywhere else, but it's a new concept in Chicago. DWM's water meter installation initiative will make customers more aware of their consumption, as well as possibly enabling a conservation pricing strategy that would increase unit cost for use above a certain level or at certain times.

Stormwater management systems reduce carbon by keeping stormwater out of the waste water treatment stream. They have the added benefit of preventing sewage contamination of Lake Michigan during heavy rain events.

Low Impact Development (LID) represents the most advanced stormwater management system technology, and has evolved from the lessons learned over the past thirty years in the United States and around the world. Simply put, LID is a new approach using decentralized integrated source control practices that make more cost-effective and efficient use of a site to maintain the watershed hydrology and water quality. Instead of the large investments in complex and costly centralized conveyance and treatment infrastructure, LID allows for the integration of treatment and management measures into urban sites. LID encourages the multifunctional cost-effective use of urban green space, buildings, landscaping, parking lots, roadways, sidewalks and various other techniques to detain, filter, treat and reduce runoff. LID is completely different from conventional management strategies. LID is flexible. It works in highly urbanized constrained areas and in environmentally sensitive areas for urban infill or retrofit projects. In a combined sewer system, LID can reduce both the number and the volume of sewer overflows.

At the building scale

One cannot manage what one does not measure. Most commercial office buildings have just one water meter for the entire building. Sub-metering sectors of use can be helpful in establishing trends and locating leaks. Supplies to HVAC equipment and toilet risers are particular areas where collected information can be useful.

©iStock/Michael Major

As noted earlier, the replacement of high-consumption toilet fixtures could single-handedly accomplish the goals of this section. Unfortunately, this represents a large capital expense for building managers as well as an inconvenience for building tenants. Most metropolitan areas have fixture retrofit rebate programs to help offset the economic and tenant inconveniences involved with this capital improvement. Chicago does not. A fixture rebate program—a sort of "cash for clunkers" for toilets—combined with a creative ad campaign could help to ease the financial burden and raise community awareness of this important initiative to achieve carbon reduction goals. An overlay district that mandates modernization with a combination of incentives and burdens should be explored. The mass purchasing power of such a program, coupled with the ability to fund it through city water bills, merits further study.

A cooling tower typically uses water at a rate of .2-.3 liters per minute per ton of refrigeration. Some of this is in the form of evaporation; the rest is used in the maintenance of water chemistry in the cooling tower basin. As water evaporates from the tower in the cooling process, chemical concentrations increase and water must be added to restore chemical concentrations to their proper balance, a process called "blowdown." Systematic monitoring and trending can help to reduce

the amount of water required to maintain water chemistry. Alternatively, a developing technology called pulse-powered water treatment eliminates the need for chemical treatment of cooling tower water.

In addition to the fact that building owners historically have been unable to overcome the capital cost hurdle of plumbing buildings with two separate waste systems, another conundrum exists for the implementation of greywater systems. The source for greywater systems is not as plentiful in a commercial building as it is in residential buildings. Without the domestic processes such as dishwashing, laundry and bathing, there's not much greywater available for recycling. However, if the capital costs can be justified, a greywater system does have the marginal owner benefit of reducing fresh water use, and the larger environmental benefit of keeping mildly contaminated water out of the waste water treatment stream. Residential occupancies are better platforms for greywater use.

A relatively new technology is the "living machine," which takes blackwater and returns it to its natural whitewater state. Microorganisms are the basis for this process.

© iStock/William Sherman

ALTERNATIVE TREATMENT PROCESSES

As strategies for carbon reduction fall into two basic categories, so too do design solutions: (1) reducing carbon load per gallon and (2) water conservation.

Generally, reducing carbon load per gallon involves minimizing or offsetting the pumping power associated with the movement of water. Most of the design solutions for this issue are engineering- based (such as improved electrical motor efficiency or variable frequency motor controllers), although the City of Chicago has demonstrated some forward thinking in the form of renewable energy arrays at one of its many pumping stations. This is a good example of a design solution that can be universally applied in some form throughout the city.

Since nearly three-fourths of the embodied carbon in a gallon of Chicago water is related to waste water treatment, close attention should be paid to this step in the water cycle. Innovative stormwater management strategies and alternative waste water treatment systems are the keys to reducing the impact of this portion of the water carbon cycle.

Natural treatment systems, ranging from building to restored wetland scales, are emerging technologies that keep water out of the mechanized waste water treatment stream. These systems have trademarked names such as Living Machines, Eco-Machines and Nutrient Farms, but they are all essentially constructed ecosystems.

An Eco-Machine can be a tank-based system traditionally housed within a greenhouse, or a combination of exterior constructed wetlands with aquatic cells inside a greenhouse. Size requirements are entirely dependent on the waste flow. It's a beautiful water garden that can be designed to provide on-site advanced treatment of blackwater.

In comparison to the waste water treatment alternatives described above, water conservation is an easy task. Sub-metering, plumbing fixture replacement and cooling tower water management are the three areas building managers and operators should focus on. Capital is the impediment to basic water conservation efforts. Public subsidies might help break this logjam.

1 Toilet fixture rebate program and ad campaign
A commercial and residential rebate program for retrofitting toilet fixtures could be launched in conjunction with an ad campaign to raise public awareness of the rebate program as well as the Chicago Climate Action Plan. Toronto's toilet rebate program includes extensive advertising to garner support and raise public awareness of the program.

2 Large-scale greywater systems
Through the pilot building surveys, a building could be identified that is in a position to retrofit a greywater recycling system. Although a residential building conversion would be convenient for this, they generally lack uses for greywater. Commercial buildings, with many more occupants and toilets, could use recycled greywater for flushing. Large-scale greywater systems suitable for commercial buildings are now available.

3 Stormwater management
Urban planning efforts should be reviewed to include preservation of open space for stormwater management. The Staten Island Bluebelt preserves natural drainage corridors including streams, ponds and other wetland areas, allowing these wetland systems to perform their functions of conveying, storing and filtering stormwater. The Bluebelts also provide important community open spaces and diverse wildlife habitats. The Bluebelt program saves tens of millions of dollars in infrastructure costs when compared to providing conventional storm sewers for the same land area.

4 Renewables at pumping stations
All pumping station locations could be analyzed and one chosen for the installation of either wind or solar renewable power arrays. A large photovoltaic array over the storage tank is planned for the Lexington pumping station.

5 Nutrient farming
To pre-empt the need for a new brick-and-mortar treatment plant for nitrogen and phosphorus in waste water, the requisite number of acres could be identified to be converted back to wetlands. In addition to removing these contaminants, the restored wetland would be a wildlife refuge that would extend the reach of the CCAP beyond the borders of the city and increase natural habitat for Illinois wildlife.

6 Biological waste water treatment
Through the pilot building surveys, a building could be identified that is in a position to be retrofitted with a biological waste water treatment system.

WASTE

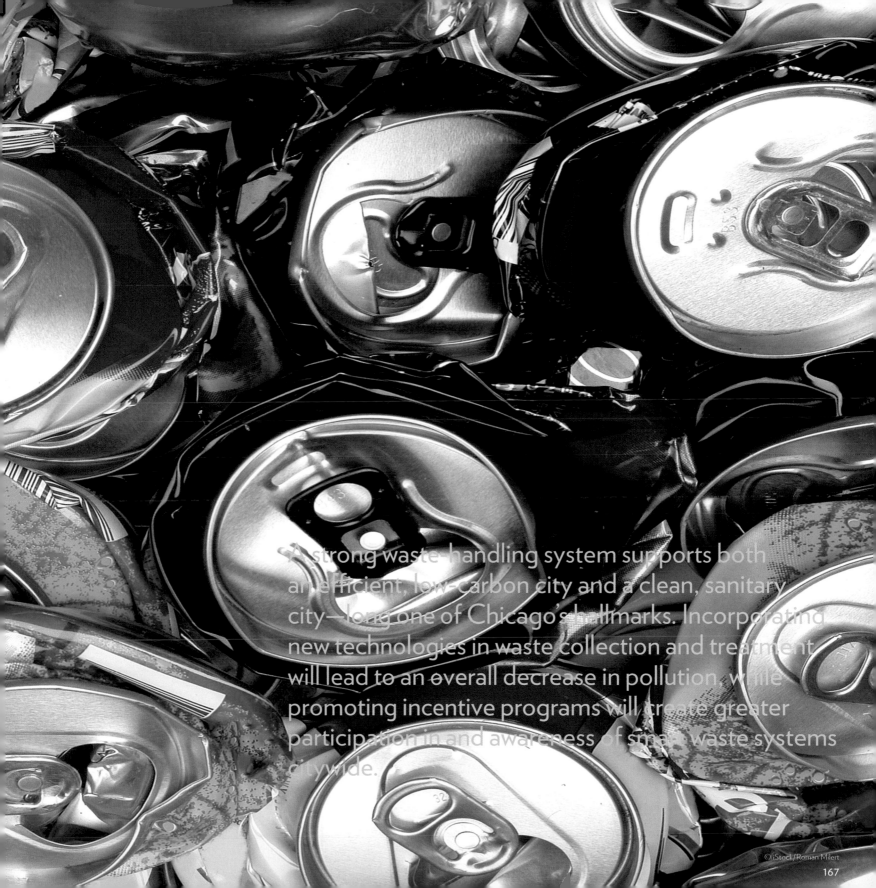

A strong waste-handling system supports both an efficient, low-carbon city and a clean, sanitary city—long one of Chicago's hallmarks. Incorporating new technologies in waste collection and treatment will lead to an overall decrease in pollution, while promoting incentive programs will create greater participation in and awareness of smart waste systems citywide.

Carbon Reduction Strategies

What if Chicago could progress beyond "reduce, reuse, recycle" to "decontaminate, decarbonize, re-energize"? A large percentage of the municipal waste generated in Chicago can be recycled or reused, but the city will need to augment its existing waste infrastructure to reach the reduction goals of the future. In the spirit of conservation, Chicago can reuse existing systems like the coal tunnels in new ways to efficiently collect and transport waste. Much of what cannot be recycled can be converted into clean new forms of energy, reducing carbon emissions even further.

In this chapter, we discuss the **Carbon Reduction Goals** based on a "reduce, reuse, recycle" strategy.

The **Analysis** section breaks down waste generation in Chicago and pinpoints which waste materials the city should focus on. Existing and new waste initiatives will help strengthen the current carbon goals set for the city, as will the use of an existing underground tunnel network to collect and channel waste out of the city.

The **Strategies** and **Precedents** sections examine pneumatic waste collection and plasma arc gasification. **Strategies** also looks at construction waste standards as a new way of collecting and disposing of waste within the city.

Finally, we identify a series of **Design Solutions and Pilot Projects**. These solutions offer a number of first steps toward turning the existing Loop building stock into a more efficient neighborhood comprised of buildings that enjoy symbiotic relationships and produce energy.

CARBON REDUCTION GOALS

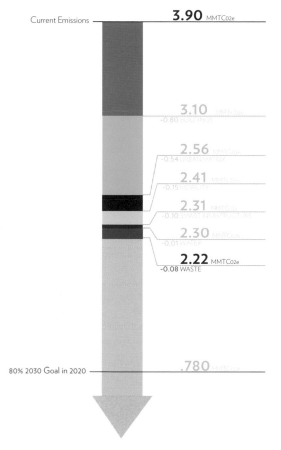

Current Emissions — **3.90** MMTC02e

3.10 MMTC02e
-0.80 BUILDINGS

2.56 MMTC02e
-0.54 URBAN MATRIX

2.41 MMTC02e
-0.15 MOBILITY

2.31 MMTC02e
-0.10 SMART INFRASTRUCTURE

2.30 MMTC02e
-0.01 WATER

2.22 MMTC02e
-0.08 WASTE

80% 2030 Goal in 2020 — .780 MMTC02e

With recycling being .07 MMTCO2e of the set objective, the recycling goal for the DeCarbonization Plan is formed by increasing the current 50% recycling quota to 85%. The process of pneumatic waste collection and other local initiatives give structure to the recycling program within the Loop target area.

As for the reuse initiative, converting waste to energy through gasification offers .01 MMTCO2e of the set goal. In the process of plasma arc gasification, waste (such as tires, biomass, river sediment, etc.) is broken down into an elemental gas. From there, the gas is converted to generate power, liquid fuels or other sustainable sources of energy.

Reduction is the third key objective. Reducing the use of virgin materials and using recycled or reused goods will strengthen the objective of the DeCarbonization Plan, with the possibility of going beyond the plan's goals.

3.4 million tons of Chicagoans' waste is left in landills every year. That is 62% of the total waste amount.

Food scraps make up 12% of waste that Americans generate each day.

Specific goals as related to waste within the DeCarbonization Plan

CHICAGO WASTE GENERATION

The aluminum can is the most valuable container to recycle. It can be recycled over and over and used to create new products without degrading its quality.

City of Chicago waste disposal for 2007, based on information from the Chicago Department of Environment's Waste Characterization Study

Decision support system (DSS)
—Collected residential

1,091,583 TONS

Industrial, commercial and institutional (ICI)

1,075,545 TONS

Construction and demolition (C+D)

2,158,080 TONS

Multi-family residential

465,705 TONS

CURRENT WASTE INITIATIVES IN CHICAGO

For the past several years, the City of Chicago has taken a progressive lead in creating waste initiative programs. Whether through recycling, reducing or reusing, these programs have moved the community forward—but the city will need to increase the standards of these programs to achieve the 2020 and 2030 carbon reduction goals.

Current initiatives
- Chicago waste-to-profit network
- Construction and demolition recycling ordinance
- The Re-Building Exchange
- Blue Cart separate collection program
- Materials exchange website
- Composting program

Outstanding issues
- The Chicago Climate Action Plan calls for a 90% recycling rate. Currently the city is recycling 40-50% of its waste.
- No requirement for large residential or commercial buildings to offer recycling
- Lack of infrastructure for recycling on the street
- Residential recycling lacks incentives. For example, the garbage collection fee is included in property taxes, and recycling can require extra fees for "blue bags" or extra effort to haul materials to a central collection point.

EXISTING FRAMEWORK

Within the existing Chicago tunnel network lies the potential to move waste through the city to one waste collection site. From there, these materials would be transported by vehicle to a recycling center or other proposed waste management programs.

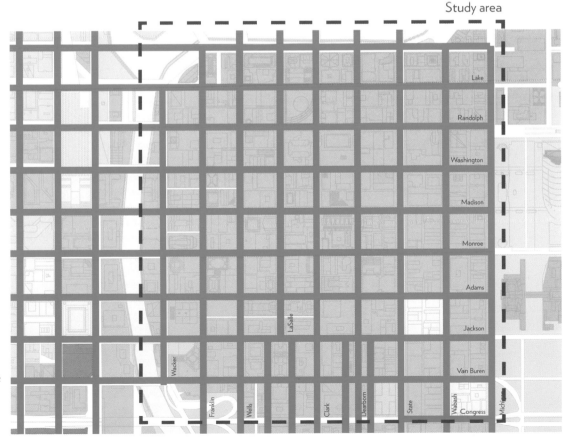

Study area

Map showing the nearly 60 miles of underground tunnels built by the **Chicago Tunnel Company in the early 1900s. Built to move coal efficiently through the city, most of the tunnels lie up to 40 feet beneath the streets of the Loop.**

PNEUMATIC WASTE COLLECTION

Pneumatic waste collection systems consist of an integrated collection station, piping system and discharge valves that are situated below vertical chutes outside all community and civic areas within a city. Waste of one type is deposited into the garbage chute, where it's stored above a discharge valve between the emptying cycles. There is a main pipe network under the valves that connects all valves and transports the waste to the collection station. These systems work best in a radius of about a mile, which is ideal for the Loop.

© www.envacgroup.com

System benefits

· A single operator for each collection station is required.
· Fewer garbage trucks leads to a reduction in CO_2 emissions, traffic pollution and noise pollution.
· No chilled garbage rooms are required.
· Lower cost is associated with annual replacement and general maintenance of the system.
· Containers are only emptied when full, reducing haulage costs and pollution.

Chicago coal tunnels

The existing infrastructure of the Loop's coal tunnels could be used for pneumatic waste collection. This system of tunnels is already connected to virtually every street and building in the Loop due to its original functions: delivering coal and mail and collecting garbage from Loop buildings before the invention of automobiles.

Pneumatic waste collectors in Stockholm, Sweden

Financing

A large overlay district for pneumatic waste collection and disposal should be explored for the study area. Its funding resources could include federal incentives encouraging recycling or waste to energy, tax increment financing and special service area financing. Refer to the Funding chapter for further details.

Life-cycle cost and benefits

As illustrated in the above diagram, piloting programs such as pneumatic waste collection can improve the overall potential for cost reduction in waste collection. The current system, which uses 500 waste-hauling trucks, is very costly compared to the operating cost of a mechanized system. Disadvantages of the current system include truck maintenance costs, labor costs and the noise and pollution the trucks create when driving on city streets. Pneumatic waste collection offers a cleaner solution that can alleviate traffic congestion and encourage recycling.

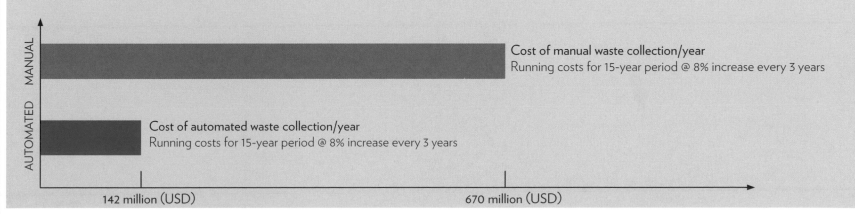

Cost of manual waste collection/year
Running costs for 15-year period @ 8% increase every 3 years

Cost of automated waste collection/year
Running costs for 15-year period @ 8% increase every 3 years

142 million (USD)   670 million (USD)

PLASMA ARC GASIFICATION

In the plasma arc gasification process, temperatures nearly equal to those on the sun's surface are used to break down the molecular structure of any carbon-containing materials—such as municipal solid waste (MSW), tires, hazardous waste, biomass, river sediment, coal and petroleum coke—and convert them into synthesis gas that can be used to generate power, liquid fuels or other sustainable sources of energy.

Gasification occurs in an oxygen-starved environment so feedstocks are vaporized, not incinerated, due to the high operating temperatures in the plasma gasification process.

- No bottom ash or fly ash that requires treatment or landfill disposal is generated in the process.

- Metals and non-combustible inorganics are melted and captured in an environmentally benign slag, which can be used as construction aggregate.

- Each plasma gasification application will have a differing environmental profile, but in general terms a plasma gasification facility will have very low emissions of NOx, SOx, dioxins and furans.

Benefits include:

- There are no emissions from the conversion of garbage to synthetic fuel gas, construction aggregate, salt, sulphur and clean water.

- Heavy metals in the waste stream are recovered and sent for safe disposal (less than 2 kg for each ton of waste processed).

- Two tons of greenhouse gas are eliminated for each ton of waste processed, resulting from a combination of displacing coal-fired electricity and diverting waste from landfills where it would produce methane, a potent greenhouse gas.

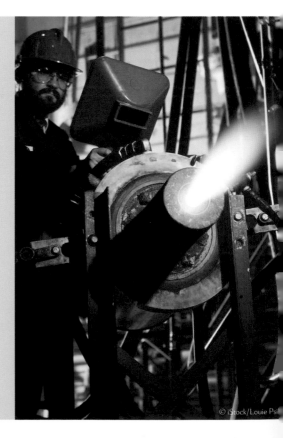

Plasma arc torch

© iStock/Louie Psi

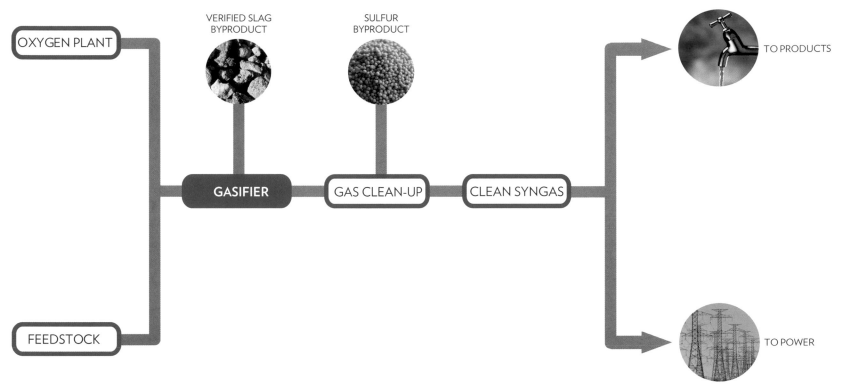

OXYGEN PLANT

VERIFIED SLAG
BYPRODUCT

SULFUR
BYPRODUCT

GASIFIER → GAS CLEAN-UP → CLEAN SYNGAS

TO PRODUCTS

FEEDSTOCK

TO POWER

Funding

Plasma arc gasification presents a good
candidate for a project finance opportunity
that involves a number of public and private
partnerships. These include a waste stream that
is diverse and comprehensive, a concession
agreement that allows that waste stream to
become feedstock that creates only byproducts
that are useful or valuable, and a plan of financing
that involves an array of federal, state and
municipal incentives and programs such as solid
waste tax-exempt bonds, renewable energy tax
credits and special service areas, among others.
Refer to the Funding chapter for more detail.

Life-cycle cost

The long-term cost of plasma arc gasification
offers significant benefits over traditional
landfills. First of all, landfills require land area
that must be bought. They also cause pollution
in the form of methane, which can be 30 times
as potent as pollution from coal-based carbon
emissions. Over the lifespan of a landfill, its
environmental cost in the form of pollution, as
well as the money required for maintenance,
can far exceed the admittedly high initial cost of
plasma arc gasification. Therefore it's important
to develop this new technology to make it a
viable option for widespread use in the future.

REDUCING CONSTRUCTION WASTE

According to the Chicago Department of Environment's Waste Characterization Study, construction waste accounts for 40% of Chicago's total municipal solid waste. Most construction waste goes into landfills, increasing the burden on landfill loading and operation. Waste from sources such as solvents or chemically treated wood can result in soil and water pollution.

Some materials can be recycled directly into the same product for reuse. Others can be reconstituted into other usable products. Many construction waste materials that are still usable can be donated to non-profit organizations. This keeps the material out of landfills and supports a good cause.

The most important step for recycling construction waste is on-site separation. Initially, this will take some extra effort and training of construction personnel. Once separation habits are established, on-site separation can be done at little or no additional cost.

The initial step in a construction waste reduction strategy is good planning. Design should be based on standard sizes and materials should be ordered accurately. Additionally, using high quality materials such as engineered products reduces rejects. This approach can reduce the amount of material that needs to be recycled and bolster profitability and economy for the builder and the customer.

Strategies

· Modularity of systems
Today's interior designers are able to create spaces, especially for office buildings with high rates of renovation or "tenant churn," that are easily adjustable. In this way, when a new tenant moves in, it's possible to move entire walls, lights, cubicles and other elements to save on construction waste.

· Re-use of structural and fit-out materials
Structural materials can be reused when buildings are renovated. Other fit-out materials, such as wood paneling, ceramic tile, carpet, doors and frames, can also often be reused. This is a growing trend in renovations that allows owners to achieve LEED points.

· Structure as finish
Another growing trend is for structural materials to be used as finishes. For example, ductwork in the ceiling can be exposed underneath beams, and concrete floor systems can be used as finish materials. This prevents the need for finish materials that are high in embodied carbon and often end up in landfills.

· Durable materials
Durable finish materials should be used in order to ensure that the materials last, reducing waste from renovation.

· Closed-loop manufacturing
Many manufacturers now offer programs whereby they recycle their own products, such as carpet reclamation.

· Post-consumer materials
The use of post-consumer recyclables in today's construction products is increasing.

PNEUMATIC WASTE COLLECTION

Roosevelt Island

Roosevelt Island has the largest pneumatic waste
system in the United States, able to collect trash from
over 20,000 people. Its pipes run at speeds of 55 mph
to an AVAC complex built in 1975. This method
of disposal minimizes air pollution and smell and is
economic in terms of manpower. Only two to three
workers run the entire plant.

**Roosevelt Island is a narrow island in the East River of New
York City between Manhattan and Queens.**

The Magic Kingdom, Walt Disney World

The AVAC system under the Magic Kingdom
moves garbage through pipes at 60 mph using
compressed air. The system operates at intervals
of about 20 minutes, taking trash from the park to a
central location where it's processed and recycled.
It was the first such system installed in the western
hemisphere.

**Disney World's Magic Kingdom saw an estimated 17 million
visitors in 2008, making it the world's most visited theme park.**

PLASMA ARC GASIFICATION

Ottawa, Ontario, Canada
The Ottawa City Council has agreed to build
and regulate a facility to process 85 tons per day
of unsorted municipal solid waste. It will generate
enough electricity to power the entire process and
about 3,600 Ottawa households. The facility is clean,
safe and critical in the fight against climate change.

Ottawa, Ontario is building a waste processing facility that will generate enough electricity to power 3,600 households.

St. Lucie, Florida
Geoplasma is constructing the first U.S. plasma
refuse plant in St. Lucie County, Florida. When the
plant opens in 2011, enough energy will be produced
to power 50,000 homes. This plant will process all of
the incoming waste for the county and begin to mine
the existing landfill for waste. The county researched
bioreactors, incineration, standard gasification,
pyrolysis and other thermal conversion technology
processes before deciding plasma arc gasification
was the most economically and environmentally
viable solution. The facility, which will be owned and
operated by Geoplasma, will cost approximately
$200 million. The state of Florida has allocated $160
million in non-taxable bonds for the project. The rest
is being financed with equity, non-taxable and taxable
bonds.

The U.S. Census Bureau 2005 estimate for St. Lucie County, Florida, was 241,305.

DECARBONIZED WASTE PILOT PROJECT— PNEUMATIC WASTE COLLECTION WITH PLASMA ARC GASIFICATION AND AGGRESSIVE RECYCLING PROGRAM

Recycling involves processing used materials into new products to prevent waste of potentially useful materials, reduce the consumption of fresh raw materials, reduce energy use, reduce air and water pollution by reducing the need for conventional waste disposal, and lower greenhouse gas emissions.

In contrast to recycling, reuse is where an item is used again for the same function. New life reuse is where the item is used for a completely new function. Historically, financial motivation was one of the main drivers for reuse.

Plasma arc gasification is a waste treatment technology that uses electrical energy and the high temperatures created by an electrical arc gasifier. This arc breaks down waste primarily into elemental gas and solid waste (slag) in a device called a plasma converter.

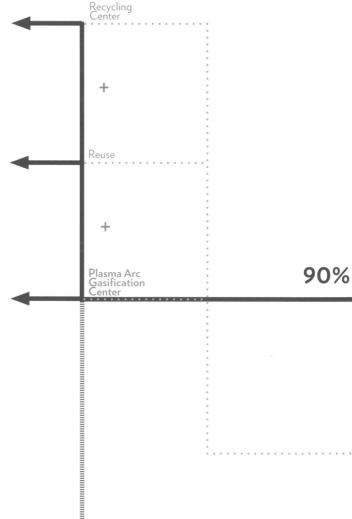

Recycling Center

Reuse

Plasma Arc Gasification Center

90%

Landfills are the oldest form of waste treatment and the most common method used around the world. Problems with landfills include their emissions of gases such as methane, undesirable fumes and the land area they occupy.

Landfill

10%

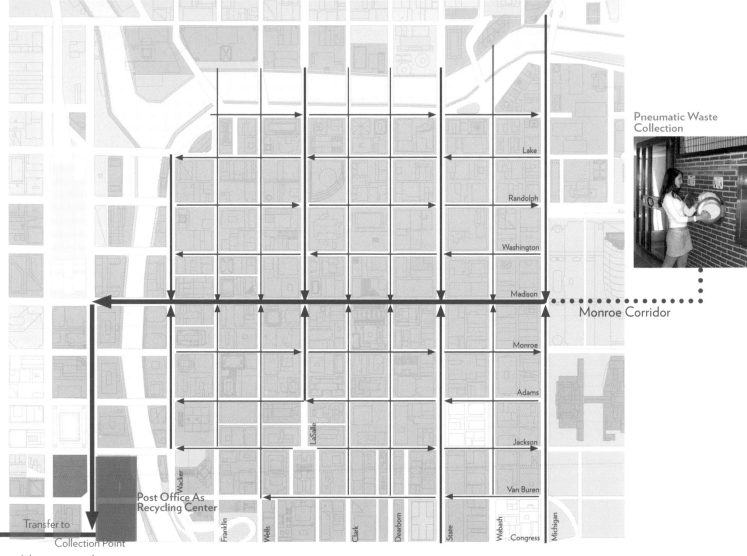

Pneumatic Waste Collection

Monroe Corridor

Lake

Randolph

Washington

Madison

Monroe

Adams

Jackson

Van Buren

Congress

Michigan

Wabash

State

Dearborn

Clark

LaSalle

Wells

Franklin

Wacker

Post Office As Recycling Center

Transfer to Collection Point

Monroe corridor

· System could be implemented in phases to test feasibility.
· Phase 1 would run along Monroe Street using existing underground tunnels.
· Phase 2 would expand the system over the study area to connect with Phase 1 tunnels ('all streets in study area except Congress currently have tunnels).
· System would transport waste to Old Post Office collection site for vehicular transport to recycling centers.

Construction and demolition waste

Construction and demolition waste consists of unwanted material produced directly or incidentally by the construction or demolition of a building. This includes materials such as brick, concrete wood, glass, insulation, nails, electrical wiring and rebar, as well as waste originating from site preparation such as dredging materials, tree stumps and rubble.

ADDITIONAL WASTE INITIATIVES

While primary strategies for waste reduction in the Loop focus on a pneumatic waste collection system and a plasma arc gasification plant, several other opportunities exist to push Chicago toward decarbonization by waste reduction. To promote a cohesive and community-supported waste reduction plan, Chicago will need to re-brand and market the strategy as a collection of initiatives, including the following proposals.

CTA credit for recycling

- Pair various city services (such as the CTA) with recycling programs.
- System works by giving incentives in the form of CTA credits for depositing a recyclable article into a nearby bin.

Pay-as-you-throw programs

- Encourage the reduction of household waste by asking residents to pay for garbage removal based on what they actually generate.
- Since garbage is paid for on a per-unit basis, residents who conserve, recycle and reuse are no longer asked to subsidize the cost of those who generate a large amount of waste.
- Current pay-as-you-throw programs report a 15-25% reduction in the amount of waste disposed and triple the amount of recycling.
- The reduction of waste would also decrease the cost of overall city MSW collection fees.

City-wide fee on plastic bags

- Require a fee to discourage the use of plastic bags.
- Fee will not only reduce waste but will become a new source of government revenue.
- Cities such as New York and Seattle have already implemented this initiative and are currently using the profits generated to fund other environmental programs.

New policy for residential recycling

- The city is currently piloting a new program for residential recycling that includes separate collection bins. This is highly recommended in order to reinforce trust with residents that their recyclables are actually being separated (as opposed to the current program, in which the trash truck typically collects blue recycling bags). Along with the pay-as-you-throw program, this will significantly increase recycling among residents.

COMMUNITY ENGAGEMENT

Chicagoans have an abundance of civic pride. To its inhabitants, Chicago is second to none. Residents must work together to encourage every single individual to take an active role in Chicago's sustainable renaissance.

© iStock/Nikada

Carbon Reduction Strategies

What if Chicagoans could set a new example for sustainable action in cities across the globe? Participation in the activities of the community enhance shared feelings of citizenship and pride. The expansion of social networks via new technologies will strengthen and change both identification and interaction among fellow Chicagoans. Citizens can leverage new social media to create larger communities dedicated to sustainability. They can also refocus their energies to develop action-oriented groups that provide incentives for more sustainable attitudes in their everyday lives.

In this chapter, we set **Carbon Reduction Goals** by estimating individual behavioral changes needed for community engagement. We then use these numbers to calculate goals for the Chicago Central Area DeCarbonization Plan.

The **Strategies** within this chapter briefly examine the concepts of branding, green team organizations, multilingual advertising and educational materials and social marketing programs. By promoting change, Chicagoans will be aware and ultimately involved in the carbon reduction initiative.

With **Precedents** such as the Willis Sustainable Technology Center, educational programs, public art, corporate programs and consumer incentives, this section identifies examples of how to involve and educate a community about sustainability and green issues.

Finally, we identify a series of **Design Solutions and Pilot Projects**. These projects, from rooftop agriculture to large-scale learning environments, can inspire and educate members of the community to incorporate sustainable ideals into their everyday lives.

CARBON REDUCTION GOALS

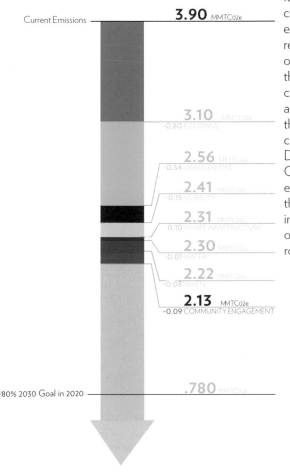

Current Emissions

3.90 MMTCO2e

3.10 MMTCO2e
-0.80 BUILDINGS

2.56 MMTCO2e
-0.54 URBAN MATRIX

2.41 MMTCO2e
-0.15 MOBILITY

2.31 MMTCO2e
-0.10 SMART INFRASTRUCTURE

2.30 MMTCO2e
-0.01 WATER

2.22 MMTCO2e
-0.08 WASTE

2.13 MMTCO2e
-0.09 COMMUNITY ENGAGEMENT

80% 2030 Goal in 2020

.780 MMTCO2e

Specific goals related to Community Engagement emission reduction within the Chicago Central Area DeCarbonization Plan

It's estimated that individual behavioral changes due to awareness of carbon emissions and climate change can result in an overall use reduction of up to 20%. The savings goal for this chapter, based on behavioral changes, is conservatively estimated at 3%, or .9 MMTCO2e. Without the engagement and support of the community, none of the goals of the DeCarbonization Plan and 2030 Challenge would be achievable. The engagement strategies must recognize the diversity of the populations who live in, work in and visit the Loop. The City of Chicago's administration also plays a role as initiator and role model.

By exposing inhabitants to a sustainable lifestyle, a sense of environmental stewardship can be created. Education will move an eager community toward change, at work and at home.

Often, community engagement is not given the weight it deserves in ensuring the successful implementation of a plan. Informed engagement matters. Inhabitants must be involved at the earliest possible phase of a project so that they can participate in an informed, expanded and stakeholder-based discussion about the potential of their community. In the process, they must be able to share their own ideas, needs and wishes.

Involving children in decision-making ensures continuing lifestyle changes.

Incentive programs are high-profile motivators for participation in sustainability programs.

Every year, 1.5 million tourists are influenced by their visits to Chicago.

© iStock/Palto

STRATEGIES

These strategies, along with effective management and governance, are key to the successful implementation of the DeCarbonization Plan.

CARBON CONSCIOUS CHICAGO

- Create a carbon reduction brand for all programs promoted by the city. This identifies the city as supporting and monitoring the impacts of the carbon reduction strategies, from creation of a new bicycle station to support for a building pursuing energy-saving strategies.
- Develop a "Green Team" supported and organized by the city. The team would consist of varied members from within schools, businesses, residential and community organizations.
- Create multilingual advertising and educational materials that can reach individuals of all ages and education levels.
- Develop a presence on social media platforms such as Facebook and MySpace to support and promote all initiatives.

WILLIS SUSTAINABLE TECHNOLOGY LEARNING CENTER

Willis Tower is comprised of about 4.5 million square feet of space, equivalent to 16 city blocks.

The tower contains 225,500 tons of building material including glass, steel and concrete.

The plumbing stretches for 25 miles.

The electricity use is over 150 million kWh per year, enough to power more than 9,000 Chicago homes.

There are 104 elevators that carry more than 1 million employees and visitors each year.

Willis Tower uses about 100,000 lights, a significant percentage of the energy load.

Willis Tower has the potential to save up to 72 million pounds of carbon emissions annually through building improvements and renewable technologies.

The Sustainable Technology Learning Center, designed for Willis Tower, is intended to be an interactive exhibit where building owners and operators, scientists, visitors and school children can learn about AS+GG's Willis Tower Greening and Modernization project firsthand. The design for the learning center includes a real-time "control room" in which energy use is tracked. There are also demonstrations of the upgrades of the exterior wall, lighting controls and mechanical systems, as well as the addition of solar panels, wind turbines and fuel cells.

EDUCATIONAL PROGRAMS

Sustainable development requires much broader public awareness and understanding of the natural resource and economic challenges facing the world in the 21st century. The 3,000 institutions of higher education in the United States are significant but largely overlooked leverage points in the transition to a sustainable world. Not only do they prepare students who will become teachers and leaders in the educational field, they also educate the students who will become leaders in other fields. These institutions also influence their alumni, many of whom constitute our nation's current leaders. The University of Chicago, Northwestern University, the University of Illinois at Chicago and Illinois Institute of Technology help make Chicago an important center of higher education.

Faculty members can play a strong role in education, research, policy development, information exchange and community outreach. They can contribute new ideas, engage in bold experimentation and contribute to new knowledge. Institutions of higher learning should place a greater emphasis on interdisciplinary, systemic and strategic ways of thinking.

Students, parents, alumni, prospective employers, organizations that fund research and education (government, industry and foundations) and the public are all consumers, clients or supporters of higher education's services. Individually, they have varying degrees of influence on academic direction and programs; collectively, they have great potential to encourage innovation in education.

1 **Bucknell University, in Lewisburg, Pennsylvania, is an active participant in the Sustainable Energy Funds Solar Scholars Program. It maintains three photovoltaic arrays that serve as a power source for the Bucknell University Environmental Center. Bucknell also hosts Solar Scholars educational workshops.**

2 **Great Seneca Creek Elementary School, in Germantown, Maryland, is the first public school in Maryland to be recognized as "eco-friendly" with a LEED Gold Certification.**

3 **The Great Lakes Science Center, in Cleveland, Ohio, is one of America's largest interactive science museums, with more than 400 "hands-on" exhibits including an iconic 150-foot wind turbine and a 300-foot solar array canopy. Combined, these energy resources provide 6% of the Science Center's daily power supply.**

Wikipedia/Derek Owen

PUBLIC ART

Public art has the power to improve our quality of life because it makes us stop and open our eyes. Think of the influence that artists have through public art projects. Public art projects often get a lot of attention. The citywide "Cows on Parade" installation was a tremendously successful tourism promotion for Chicago, producing an economic impact estimated at $200 million. More than a million visitors flocked to Chicago to see the cows. For public art projects, going green does not create excessive work or expense. Why not package a message of sustainability with the art?

4 Adrian Smith + Gordon Gill Architecture was proud to be among the 100 artists contributing to Chicago's Cool Globes public art program.

The Cool Globes exhibit featured more than 100 sculpted globes, each five feet in diameter, displayed along the lakefront from the Field Museum north to Navy Pier. Each globe is designed to inspire individuals and organizations to take action against global warming.

5 The Crown Fountain, designed by Spanish artist Jaume Plensa and inspired by the people of Chicago, consists of two 50-foot glass block towers at each end of a shallow reflecting pool. Faces of Chicago citizens are projected on LED screens. Water flows through an outlet in the screen to give the illusion of water spouting from their mouths. The faces were taken from a cross-section of 1,000 residents to represent the racial and ethnic diversity of the city. An anemometer at the top of each tower monitors the wind speed, reducing pump speed and flow as the wind picks up.

CORPORATE PROGRAMS

Corporate community engagement is a company's approach to its involvement in the community in which it operates. Companies, in partnership with the public and non-profit sectors, invest in local communities to create healthy and sustainable circumstances in which they can continue to operate well. Corporate community engagement programs are designed to meet community needs as well as core business objectives. This kind of engagement is not about simple check-writing. It's about positive publicity and environmental stewardship.

Effective community engagement combines financial support with:
- The deployment of the skills, talents, energy and enthusiasm of employees, managers and key others
- Gifts in kind (including company products)
- The use of business premises and facilities
- The transfer of skills, expertise and access to business contacts and networks

Corporate community engagement programs can take many different forms, but all forms seek to bring tangible and sustainable benefits to communities. The potential for company benefits is increasingly well-known. Corporate community engagement provides a unique opportunity for active, experiential learning about other sectors and situations, as well as learning about oneself.

1 Bank of America has made a sustained commitment to America's communities, pledging $750 billion in loans and investments for community development over the next 10 years. Among its priorities are affordable housing, economic development and urban redevelopment projects, many with a strong sustainable component.

2 Since its founding in 1995, Stationnement de Montreal has supported the fight to reduce the high school dropout rate by giving a portion of its profits to the Fonds Ville-Marie. In 2007, when the company was asked by the City of Montreal to implement a public bike system, Bixi Bike of Montreal, it decided to continue supporting this cause as well.

INCENTIVES

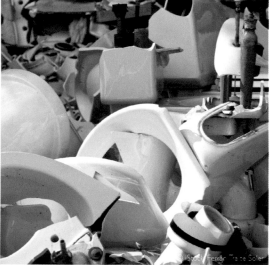

Incentives can be used to motivate people to engage in conservation activities, and to reward them for engaging in such activities. Incentives can assist both groups and individuals. For example, groups can be motivated to continue working on environmental projects through financial or in-kind administrative support, while landholders can receive tax rebates for positive land management. Projects, policies and plans will often lead to more effective outcomes if they're linked to incentive mechanisms.

Incentives can include:
- Financial incentive mechanisms (rate rebates and grants, land acquisition, in-kind support)
- Non-financial motivational incentives (local awards, community recognition, training, technical support)
- Planning or development incentives (tradeable or transferable development rights, compensatory clauses, development bonuses, negotiating for conservation gain)
- Property right mechanisms (management agreements and revolving funds)
- Revenue-raising mechanisms (environmental levies and developer contributions)

Not everyone will be influenced by incentives, and in some cases different strategies may be more efficient and effective. To be truly effective, local level incentives should also be supported by regional, state and national programs.

To allow individuals to select the appropriate level of involvement for their circumstances, a range of incentives should be available. It's often necessary to start with small steps and continue to work over time to increase the level of commitment.

In addition to rebates for high consumption plumbing fixtures, a Toronto rebate program worked in concert with an ad campaign to raise public awareness of the program and encourage participation.

CHICAGO ECO-BRIDGE

The Chicago Eco-Bridge by AS+GG would complete the last major recommendation of Daniel Burnham's 1909 *Plan of Chicago* in a thoroughly modern way that celebrates the city's past and future.

Courtesy of the Art Institute of Chicago

Illustration for Daniel Burnham's 1909 *Plan of Chicago*

The two-mile bridge, a breakwater in the Monroe harbor, would celebrate Chicago's position as the greenest city in the United States. The bridge would create a grand new civic space, providing recreational opportunities and offering unparalleled views of the skyline from a central observation tower.

The central Eco-Tower, designed to harvest wind and solar power, would function as a public observatory and also include food service and/or other amenities. A plaza at the tower site could be used as a temporary or permanent performance area. At either end of the bridge, interactive exhibit spaces could house a Great Lakes Museum or other cultural facility to complement the nearby Museum Campus.

Vertical-axis wind turbines incorporated along the bridge would add economic value by producing clean, renewable energy and showcasing Chicago's dedication to sustainability. The turbines double

as pieces of sculpture, offering an opportunity for collaboration with visual artists and engineers, and the creation of a lively public art walk. The turbines could also be outfitted with LED lights, creating a soft necklace of light surrounding the harbor in the evening.

The Eco-Bridge would be easily accessible from Grant Park and the downtown area. Pedestrian walkways and dedicated bicycle paths would encourage healthy and sustainable modes of transit. An electric or solar trolley system could further augment the bridge's accessibility, providing transport from neighboring downtown areas.

The Eco-Bridge would provide a haven for fish, wildlife and water plants, showcasing the ecology of the Great Lakes. It could also become a freshwater and Great Lakes study center.

© iStock/Francesco Fiondella

CHICAGO ECO-BRIDGE

Sculptural wind turbines

Green, pervious paving along trolley and pedestrian paths

Biking/running pathways

Trolley route

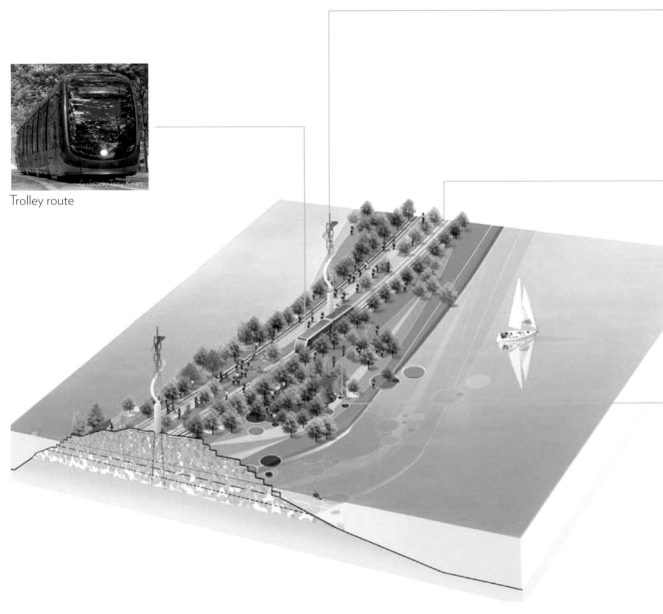

CITY HALL ROOFTOP VEGETABLE GARDEN

Chicago City Hall and its administrators play a key role as role models to the citizens of the city. City Hall currently has a very successful green roof that is now in its tenth year. It's proposed that its use be expanded as an organic garden whose produce could be sold at the marketplaces in the Loop. This would serve as an educational example of urban agriculture.

The DeCarbonization Plan seeks to capitalize on the current movement toward growing and harvesting local food. Using part of the existing City Hall green roof, or approaching the County Building side of the structure to use their space for growing fruits and vegetables, could really promote this concept.

The produce could be given to local food banks and shelters or sold at the farmers' market in Daley Plaza, or at local restaurants, with sales donated to charity.

A culinary program that focuses on healthy, locally sourced food and uses produce grown on-site as its key ingredients, encouraging both healthy and low-carbon cuisine, could be developed to further promote and advertise the concept of urban agriculture.

Reducing packaging of processed foods is another key factor in growing local produce that could be part of an education program on reducing waste.

Chicago farmers' markets can provide free resources and educational opportunities about local sustainable issues to shoppers and farmers.

ENVIRONMENTAL LEARNING SCHOOL

The education of children is a key factor in developing ongoing sustainability. As part of the DeCarbonization Plan, the Loop will increase in density, with the addition of quality schools serving as a key part of the strategy to entice young families to live here. We propose that a magnet school or group of schools with a special focus on environmental learning be created in the heart of the Loop.

These schools will have programs that focus on experiential and project-based fieldwork, creating different methods of learning. They will integrate scientific inquiry, technology and the arts through a stimulating cultural and natural environment, thereby teaching children that learning about science doesn't have to be limited to textbooks. The schools can incorporate learning labs such as rooftop Living Machines that also naturally treat the water used in their buildings.

Existing buildings can be reused with rooftop playgrounds for outdoor recreation.

These approaches to learning will encourage environmental stewardship through daily exposure to responsible living.

Educational opportunities abound as the Living Machine becomes a Living Classroom.

© iStock/Acky Yeung

THE GREEN CITY

Burnham's *Plan of Chicago* was introduced in 1909. The key ideas of the *Plan* are a belief in mass education, the efficiency movement, the efficacy of public relations in implementing public planning and the influence of the urban environment on human behavior. We propose that a new textbook called *The Green City* be written as a summary of Chicago's vision for a carbon-free future and taught as part of the curriculum at Chicago public schools.

In 1920, the *Plan of Chicago* was introduced to eighth- and ninth-graders in the city's public schools through *Wacker's Manual of the Plan of Chicago*, the work of Charles H. Wacker, chair of the City Plan Commission, and author Walter D. Moody.

The *Manual* was developed to help sell Burnham's vision to the public. It explained the benefits of the Plan and introduced the profound idea that citizens have a responsibility for the stewardship of a great city. It taught that small individual contributions and decisions ultimately have the greatest impact on a city's success.

The Green City textbook would be a primer for the DeCarbonization Plan. The DeCarbonization Plan signifies the start of a new era in urban design. *The Green City* would teach that this is the age of sustainability, and that energy use and carbon reduction, along with sustainable urbanization strategies, are vital to the well-being of the inhabitants of the city. The book would explain the logistics of DeCarbonization and the intent of strategies to mitigate carbon emissions.

ENERGY

Over the course of this century, we will witness the evolution of our energy infrastructure. It will be cleaner and smarter, as well as more localized, distributed and shared. This will place greater demands on every designed element, from buildings to parking lots, but it will also create many new opportunities.

ENERGY
Carbon Reduction Strategies

What if our cities could not just consume energy, but be active participants in the energy equation? As a result of a growing population and the widespread use of the personal electronics that typify our information age, energy consumption in the United States has increased exponentially over the past several decades. This rapid growth in demand is taxing an aging U.S. energy infrastructure, leading to brownouts and lost economic output. It has also led to increased reliance on foreign energy sources, creating the potential for geopolitical tensions and consequential environmental damage.

In this chapter, we establish the **Carbon Reduction Goals** by aligning with the State Renewable Energy Portfolio Standards and the Chicago Climate Action Plan relative to the DeCarbonization Plan study area.

In **Analysis**, we study the regional trends in energy supply for the state of Illinois and for Chicago in aggregate and broken down by energy source: coal, natural gas, nuclear, hydroelectric, wind and other renewable sources. To further the analysis, these trends are projected into 2020, taking into account pending federal legislation that would mandate reductions in aggregate carbon emission factors, suggesting an increased reliance on low-carbon technologies such as wind turbines, windmills and photovoltaic panels. These systems are also compared based on their capital intensity and annual carbon abatement costs.

In the **Strategies** section, these technologies are delineated by on-site renewable and co-generation energy strategies, complemented by off-site renewable and district energy strategies.

Finally, in the **Design Solutions and Pilot Projects** section, we propose three distinct scales of solutions that integrate elements of a new energy equation with Chicago's built environment.

CARBON REDUCTION GOALS

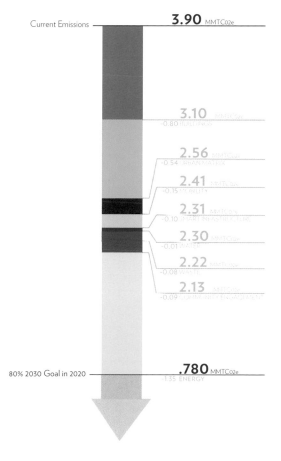

Current Emissions — **3.90** MMTC02e

3.10 MMTCO2e
-0.80 BUILDINGS

2.56 MMTCO2e
-0.54 URBAN MATRIX

2.41 MMTC02e
-0.15 MOBILITY

2.31 MMTC02e
-0.10 SMART INFRASTRUCTURE

2.30 MMTCO2e
-0.01 WATER

2.22 MMTCO2e
-0.08 WASTE

2.13 MMTC02e
-0.09 COMMUNITY ENGAGEMENT

80% 2030 Goal in 2020 — **.780** MMTC02e
-1.35 ENERGY

The equivalent of 1.35 million metric tons of carbon emissions must be eliminated to reach 2030 Challenge goals of 80% reduction in 2020.

As the 21 coal plants in Illinois reach the end of their useful life and are decommissioned, a tremendous opportunity exists to significantly reduce the carbon emissions associated with powering the city of Chicago. There are several promising technologies existing today that can be deployed both within the study area, in a building-integrated form, as well as outside the study area in an offset form.

At the state level, the transformation has already begun. In August 2007, Illinois adopted one of the most aggressive Renewable Energy Portfolio Standards in the country, requiring utilities to provide a steadily increasing percentage of their energy from renewable energy sources each year. Accordingly, both utilities and alternative retail electric suppliers must now generate at least 5% of their energy from renewable energy sources, and 25% by 2025. The new legislation also sets minimum amounts of energy that both utilities and alternative retail electric suppliers must obtain from specific energy sources. Since 2007, utilities have been required to obtain 75% of their renewable energy from wind.

The built density of the DeCarbonization Plan study area makes the integration of on-site renewable and distributed energy systems very difficult, requiring careful analysis for effective integration, but new, high-performance buildings will be equipped with both of these forms of energy. In terms of DeCarbonization solutions, energy production should be the last solution to be used. There is more value in first conserving energy wherever possible. Once all options for energy conservation have been exhausted, then on-site and off-site renewable and distributed energy generation can fulfill the remaining need.

The remaining need

To reach the 2030 Challenge goals requires a further 1.35 MMTCO2e reduction in carbon emissions. A typical 3 MW windmill, with a 35% capacity factor (a variable taking into account the unpredictability of wind), generates about 9198 MWh of energy. To account for this amount of carbon (1.35 MMTCO2e required to meet the goal) would require approximately 220 of these 3 MW windmills. The land area required for this number of windmills is roughly 54,000 acres (.4% of the surface area of Lake Michigan). Capital cost for wind energy, on a unit cost basis, is about $1.6 million/MW (turbines 76%, foundations 7%, grid connection 9%, land acquisition and control systems 8%). Two hundred and twenty windmills would have an initial pricetag of about $1.1 billion.

93% of energy consumed in the U.S. comes from non-renewable sources.

One wind turbine can power 300 homes.

Water is the world's most used renewable resource.

Enough sunlight falls on the surface of the earth every hour to power the planet for a year.

Vertical-axis wind turbines and solar panels are in use at the PepsiCo Chicago Sustainability Center.

REGIONAL ENERGY PRODUCTION AND CONSUMPTION TRENDS

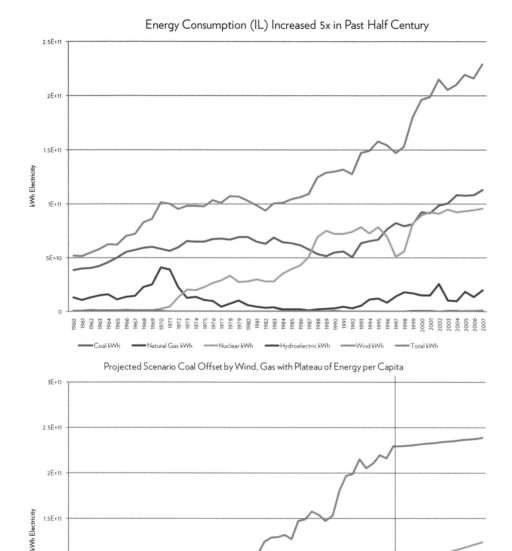

Energy Consumption (IL) Increased 5x in Past Half Century

Projected Scenario Coal Offset by Wind, Gas with Plateau of Energy per Capita

Energy production, and thus carbon emissions, has accelerated over the past half-century. Pending federal legislation to curb greenhouse gas emissions and increase the energy efficiency of systems will result in increased reliance on low-carbon energy technologies—whether traditional hydrocarbon-based sources with carbon capture and sequestration methodologies, combined heat and power (CHP) applications of hydrocarbon sources, or renewable energy such as wind and solar.

CARBON COST ABATEMENT CURVE
AND CAPITAL INTENSITY

Combined heat and power, in concert with renewable energy systems, is an excellent complement to energy-efficiency measures. Reduction in demand will reduce requisite capital expenditures for equipment and the technologies that are readily available today—unlike clean coal technologies or nuclear fusion, which are potentially decades away from full-scale deployment.

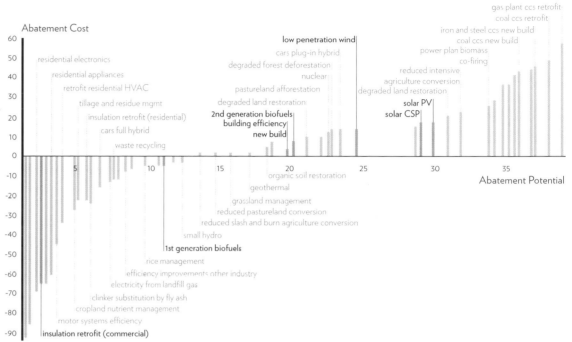

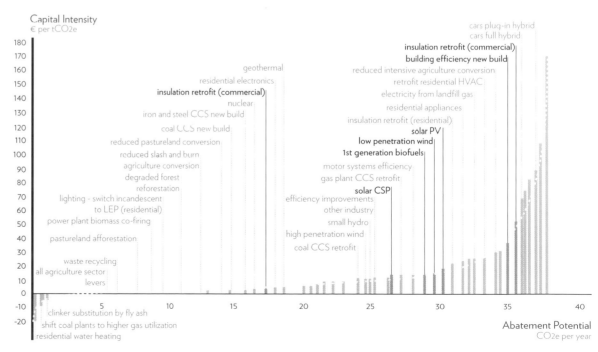

EXISTING AND FUTURE APPROACHES TO ENERGY

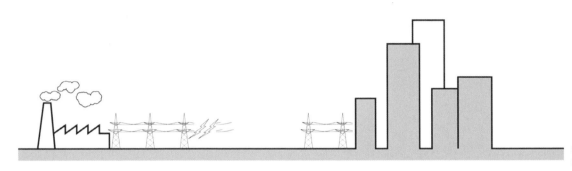

Old technology, old philosophy: An "out-of-sight, out-of-mind" philosophy, in which the environmental consequences of old thermal power plants are removed from the consumer.

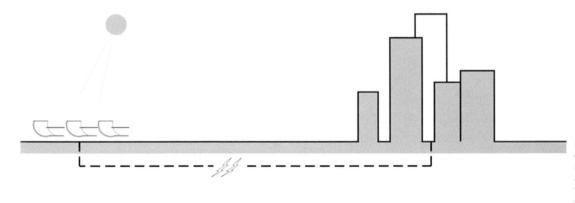

New technology, old philosophy: An evolutionary approach to energy supply, in which increased distribution losses and systemic inefficiencies limit the economic feasibility of large-scale adoption of low-carbon technologies.

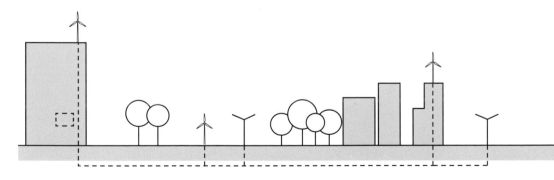

New technology, new philosophy: A revolutionary approach in which energy supply is brought to the consumer for increased efficiency and reduced infrastructure costs.

ON-SITE OFF-SITE

On-site renewable
Infrastructure costs can potentially be reduced by integrating renewable energy strategies such as wind or solar into the built environment. Possibilities include integrating photovoltaics within facades and small wind turbines on rooftops.

Off-site renewable
The state of Illinois has highly favorable conditions for large-scale wind farm deployment in the central areas of the state and Lake Michigan. Additionally, relatively high levels of sunlight make photovoltaic energy an alternative means of low-carbon power to hedge against variable weather conditions.

RENEWABLE

 Photovoltaics

 Wind

 Photovoltaics

 Geothermal

 Wind

 Wave

DISTRIBUTED

 In building co-generation

 Central cooling energy plants

On-site distributed
More than one third of the latent energy used to generate power through traditional remote thermal power plants is lost as heat. Integrating small-scale combined heat and power plants would allow for this heat to be reclaimed for building use such as domestic hot water, heating and cooling through absorption cooling.

Off-site distributed
Not all buildings are capable of readily accepting on-site power generation strategies. But there are opportunities to capitalize on economies of scale attributed to district energy systems such as cooling and heating through combined heat and power technologies.

RENEWABLE FRAMEWORK

The primary renewable energy technologies considered within this study are geothermal/river-sourced energy, wind energy, photovoltaic energy and solar thermal energy. A common characteristic of all of these technologies is their high degree of dependence on site and environment.

Funding

The Funding chapter identifies a number of federal programs that encourage the shift to renewable and distributed energy sources.

Technology	Installation Cost with Federal Incentive $/Watt	10 Year Life-time Cost $/kWh	Cost of 10 Year Carbon Offset $/kgCO2	% Yearly Reduction in CO2 for 1000kW Renewable/Distributed System for 1Msf Office
Geothermal/River-sourced Energy	$ 3.30	$ 0.06	$ 0.10	1.5%
Small wind	$ 6.20	$ 0.33	$ 0.48	3.7%
Large wind	$ 1.50	$ 0.08	$ 0.12	3.7%
Concentrated Photovoltaics	$ 7.00	$ 0.50	$ 0.71	3.2%

© Corbis/Dave G. Houser

Geothermal/River-sourced Energy

· Using this technology, the ground or a body of water is used to exchange heat with a building either directly or indirectly. Tall buildings have a limited potential for ground-sourced cooling due to limited site area. River-sourced cooling must be carefully monitored to prevent river ecology from changing. Due to the relative shallowness of Lake Michigan, deep water cooling is not viable.

Wind

· Wind generation requires a site with moderate winds (on average greater than 3-4 m/s) with limited turbulence. It has 30%-40% typical efficiency, depending on technology (Drag, Airfoil). Vertical axis models take advantage of multiple wind directions, while a horizontal axis requires a yaw mechanism. Wind effects around buildings can potentially enhance performance through acceleration.

Concentrated Photovoltaics

· CPV is much more efficient than flat panel PV and uses less silicon, but is subject to clouds and snow.

DISTRIBUTED ENERGY FRAMEWORK

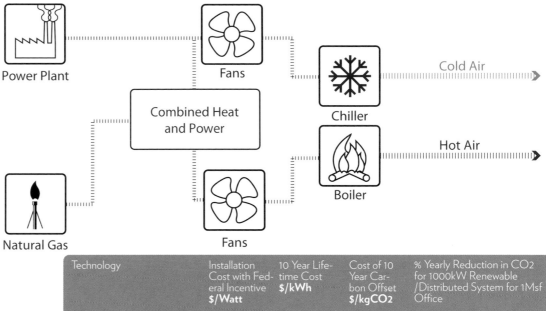

Technology	Installation Cost with Federal Incentive **$/Watt**	10 Year Lifetime Cost **$/kWh**	Cost of 10 Year Carbon Offset **$/kgCO2**	% Yearly Reduction in CO2 for 1000kW Renewable /Distributed System for 1Msf Office
Micro Turbines	$1.80	$0.01	$0.10	6.4%
Fuel Cells	$3.38	$ 0.00	$ 0.03	10.7%
Algae Farms	xxx	xxx	xxx	xxx

Micro Turbines

- Micro turbines are well suited for combined heat and power (CHP) applications. They can operate on a number of gas fuel sources and can be started quickly and easily, but must operate near peak load to realize potentially high efficiencies.

Fuel Cells

- Fuel cells can also operate on a number of different fuel types and are well suited for CHP applications. They are a highly scalable technology.

Algae Farms

- Algae is an emerging technology that generates energy through biofuel and protein. It can be produced on a mass scale in algae production farms or possibly integrated into urban building elements.

ENERGY OFFSETS FOR COMMUNITY REINVESTMENT

Large-scale, off-site renewable energy projects could take the form of community reinvestment by various firms, or a consortium of firms, as a form of corporate citizenship.

Although renewable and distributed energy sources can play a large part in any carbon reduction plan, they are often not easily integrated into existing buildings in the urban environment. Dollars earmarked for this purpose would be better spent as an investment in a larger off-site project, similar to carbon offsets, but in physical form.

Some building owners, such as banking institutions, have a legal obligation for community reinvestment. These institutions can take advantage of this opportunity to offset their fossil fuel energy use while simultaneously fulfilling their obligations. These projects can take numerous forms, depending on the capital available. Single corporations can initiate small projects or a consortium can be formed for large-scale undertakings.

Opportunities for off-site renewable energy projects range from solar thermal collectors atop low-income housing to floating photovoltaic arrays in Lake Michigan.

© iStock/Marcus Gann

© Corbis/Andrew Aitchison

BATTERY STORAGE   TO GRID

CLEAN TECHNOLOGY TOWER

Designed by Adrian Smith + Gordon Gill Architecture, Clean Technology Tower is a high-performance, net-zero-energy mixed-use development. It was designed both for a specific site in Chicago and as a prototype to demonstrate sustainable architecture principles that can be applied around the globe.

Building on the concept of biomimicry, Clean Technology Tower uses advanced sustainability systems and strategies to foster a symbiotic relationship with its environment. Sited and formed to harness natural forces, the tower refines conventional methods of capturing those forces to increase efficiency.

Wind turbines at the building's corners capture wind at its highest velocity as it accelerates around the tower. The turbines become denser as the tower ascends and wind speeds increase. At the tower's apex, where wind speeds are highest, a domed double roof cavity captures air, creating a substantial wind farm; negative pressures ventilate the interior. The dome is shaded by photovoltaic cells that capture the southern sun.

While the prototype design is specific to Chicago, its guiding principles and applied technologies are universal. Clean Technology Towers could be built around the world, and each would be distinctly formed in response to its particular site.

The Chicago design features about 1.8 million square feet of office space, 300,000 square feet of hotel space, a spa and street level retail. Office space is located on upper floors to maximize views and take advantage of natural daylight. Dedicated elevators provide access for both office and hotel tenants to all of the tower's amenities. The tower's domed top offers unrestricted skyward views, creating a grand atrium space.

AS+GG's Clean Technology Tower in Chicago (opposite page) is formed to direct wind into vertical-axis wind turbines integrated into the exterior wall (upper right) and roof (lower right).

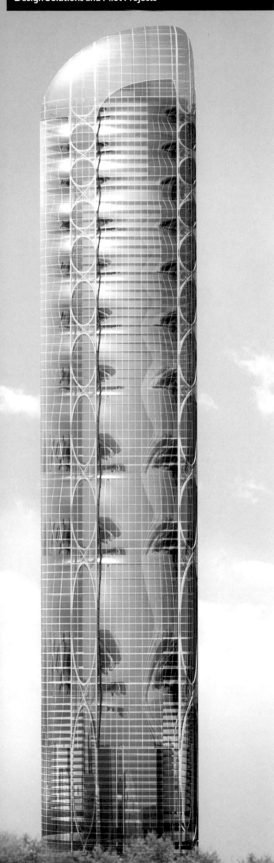

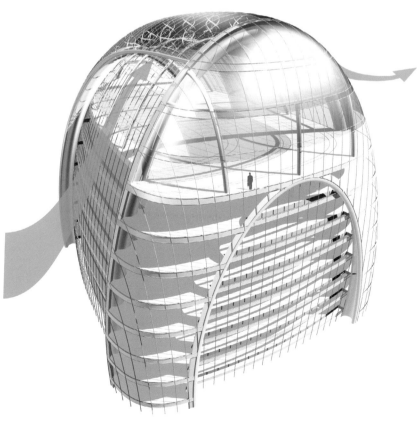

SOLAR EV DOCK

The Solar EV Dock, designed by Adrian Smith + Gordon Gill Architecture for Carbon Day Automotive, is a prototype canopy structure intended to harvest solar energy and power electric/hybrid vehicles.

The Solar EV Dock was viewed by members of the International Olympic Committee during its visit to Chicago on April 5, 2009. Built at Douglas Park, it has been relocated to Northerly Island, where it will eventually be used by the City of Chicago to fuel the city's fleet of electric/hybrid vehicles.

The Solar EV Dock features photovoltaic panels and an integrated conduit through which energy is donated to the power grid, which in turn can be tapped via an attached charge point to fuel electric cars, bikes and scooters. A stand-alone application could be developed for areas without sufficient power grid infrastructure.

The canopy is designed to work as a single structure or multiple structures linked back-to-back (in which case the Solar EV Dock creates a shaded corridor for users in the interstitial space). The canopy also creates shade for vehicles that could be charged over the course of a workday, offsetting the carbon emissions associated with commuting to work.

The Solar EV Dock can be designed to incorporate a range of photovoltaic technologies at varying orientations to accommodate installations anywhere in the world. Using the industry-standard mono-crystalline PV module, 4000-6000 kWh of carbon-free electricity is generated from a structure covering two vehicles. That is enough energy to provide up to 30,000 carbon-free miles per year, or about the amount required for a 20-mile (30-minute) work commute each way for both cars over the entire year.

In large-scale applications of the Solar EV Dock, employee parking lots could be converted into giant plug-in charging stations. Surplus energy could be donated to the power grid.

The canopy also has the potential to collect rainwater, which could irrigate adjacent agricultural areas or parklands.

During daylight hours, excess photovoltaic power is fed back into the electrical grid (opposite page). At night, the automobile battery stores power (upper right). Solar canopies provide shaded parking in addition to acting as a renewable energy source (below).

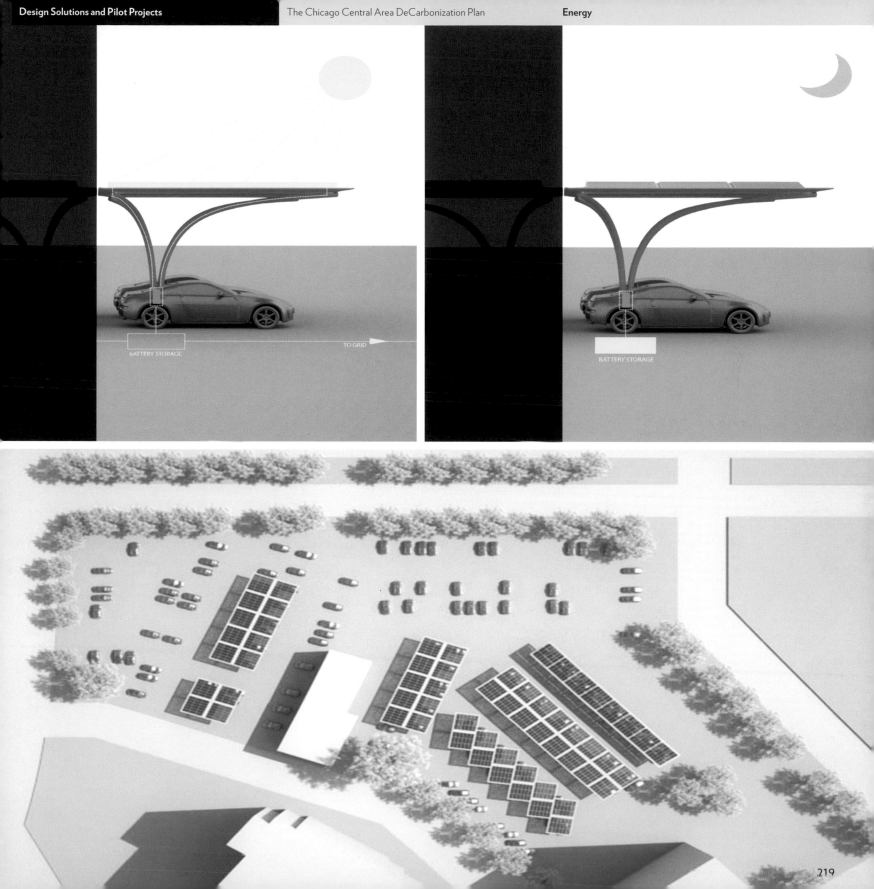

BATTERY STORAGE

TO GRID

BATTERY STORAGE

WILLIS TOWER GREENING AND MODERNIZATION

AS+GG's design for greening Willis (formerly Sears) Tower will use sustainable technologies and strategies to make the 1973 landmark much more energy efficient. The project will reduce base building electricity use by 80% (including energy savings and distributed energy in the form of fuel cell co-generation), equivalent to 68 million kWh or 150,000 barrels of oil per year—akin to the amount of energy saved by taking an entire suburb off the grid. Each 10% of energy savings at Willis is equivalent to carbon emissions from about 2,000 cars or the carbon offset from planting 10 million trees.

The greening project at Willis Tower—the tallest building in the United States and also one of the largest in terms of floor area, at 4.5 million sf—will demonstrate how to save a significant amount of energy and improve the occupant experience of an existing building during a large-scale 21st-century modernization.

A key component of the project is improving the efficiency of the building's exterior envelope and 16,000 windows, which could save up to 30% of base building energy overall and up to 50% of heating energy. The introduction of a thermal break is being investigated.

Mechanical system upgrades are planned in the form of new gas boilers that use fuel cell technologies, which generate electricity, heating and cooling at as much as 90% efficiency. Mechanical upgrades will also include new high-efficiency chillers and upgrades to the distribution system, which will be possible due to the exterior wall improvements.

The tower's 104 high-speed elevators and 15 escalators will be modernized with the latest technology to achieve a 40% reduction in their energy consumption.

Water savings will be realized with conservation initiatives through upgrades to restroom fixtures, condensation recovery systems and water-efficient landscaping, which will reduce water use by 40% and save 24 million gallons of water each year.

AS+GG has also designed a highly sustainable, 50-story five-star hotel to be built on the south side of the Willis Tower plaza. The hotel will operate using only part of the energy savings from the Willis Tower greening project, which means that the hotel will draw net zero energy from the power grid. The hotel also features integrated renewable energy such as a solar roof deck and wind turbines that are optimized in special enclosures to accelerate and control the wind.

EXTERIOR WALL
The insulation value of the exterior wall system which consists of more than 16,000 windows and metal panel, will be upgraded. These upgrades are estimated to save 11% of overall energy and 30% of heating energy.

MECHANICAL SYSTEMS
The mechanical systems will be upgraded to save energy through higher efficiency equipment, stack effect reduction and the introduction of fuel cell boiler plants.

DAYLIGHTING
By expanding daylighting harvesting, Willis Tower will save electrical energy and improve the work environment. Proposed methods include advanced lighting controls, higher ceilings, light shelves and more efficient light fixtures.

GREEN ROOFS
Green roofs at Willis Tower will be the highest in the United States, reducing storm water runoff, improving insulation, and helping to mitigate the urban heat island effect.

WATER SAVINGS
Willis Tower will upgrade its plumbing systems using the latest technology to reduce its water usage by up to 24 million gallons annually.

OPERATIONS AND MAINTENANCE
Willis Tower has introduced several new operational policies, including green cleaning, a bike-sharing program, bike storage, and a new recycling policy. The tower is also pursuing LEED certification for existing buildings.

SOLAR HOT WATER
Solar hot water panels at the 90th-story roof level will help provide water for the building's restrooms. They will also be the highest solar panels in the United States.

WIND TURBINES
Taking advantage of the tower's height and location, wind testing will be performed at the highest roofs, to help advance the use of building-integrated wind power in existing structures.

VERTICAL TRANSPORTATION
Willis Tower's elevator equipment will be modernized to provide greater energy efficiency and continued reliability.

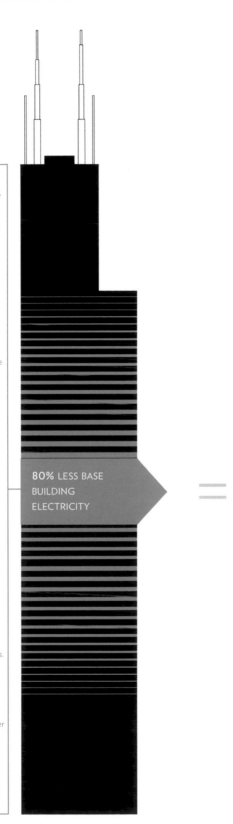

80% LESS BASE BUILDING ELECTRICITY

5,000,000 FEWER MILES OF HIGHWAY DRIVING

\+

50,000 FEWER BARRELS OF CRUDE OIL

\+

A HOTEL USING NET ZERO ENERGY FROM THE GRID

\+

ELECTRICITY FOR 2,500 AVERAGE CHICAGO HOMES

\+

5,000,00 LIGHTBULBS SAVED

WILLIS TOWER PROPOSED HOTEL

The Willis Tower project also adds an AS+GG-designed hotel to the site on the tower's south plaza. The overall concept is to add a new building to an existing site while offsetting its power use with reductions from the existing tower. In this way, the new hotel can be built with zero impact to the power grid.

The proposed hotel exemplifies the latest in sustainable design. It's intended as an icon of future green technology, sharing a site with the Willis Tower, an icon of the Chicago skyline and a great achievement for its time.

The proposed hotel features integrated wind turbines designed to receive the prevailing southwest winds. The shape of the tower is curved to streamline the wind into glass enclosures designed to accelerate the wind and house the turbines to minimize turbulence.

Skygardens are another prominent feature of the proposed building. These gardens allow greenery in the vertical setting of the city. In addition to the skygardens in the tower itself, the entire Willis Tower site is to be landscaped, creating public outdoor green space for the city. This park-like setting will mitigate the current stormwater runoff from the existing granite plaza and allow increased shading, and shielding from ground-level wind.

At the top of the hotel tower is an indoor conservatory whose roof consists of a glass skylight with integrated photovoltaics. This solar deck will redefine how roofs of highrise buildings are viewed in Chicago, setting an example of solar energy generation and offering a pleasant view from the adjacent buildings looking down.

The hotel is designed to a LEED Gold standard, and includes one of Chicago's first double-walled facades as well as efficient mechanical systems.

Willis Tower and the hotel are designed to work in harmony. Synergies between the varying building uses will be explored to further energy efficiencies where possible.

The Willis Tower proposed hotel (1) takes full advantage of synergies between day and night operations, (2) uses building form to funnel wind into integrated turbines, (3) features a solar deck and (4) skygardens with breathtaking views.

1

Night
5 p.m. - 9 a.m.
Peak System Use

Day
9 a.m. - 5 p.m.
Peak System Use

A Symbiotic Relationship

2

3

4

PARAMETRIC MODEL

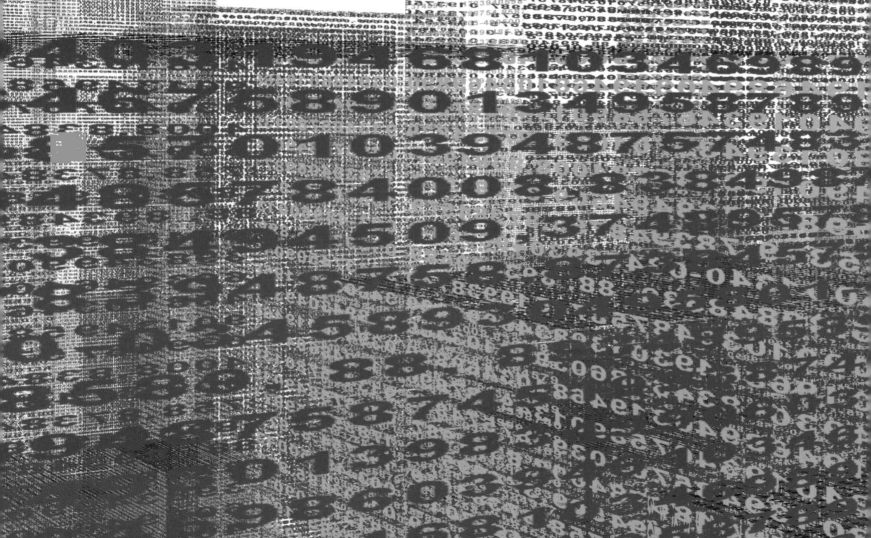

Given the mathematical challenge of multiple variables, unknowns and criteria, the DeCarbonization Plan's sustainable planning processes are complex. New computational tools are needed to simulate and predict the dynamic effects and interrelationships of a vast number of possible strategies. To this end, AS+GG developed a Parametric Model.

PARAMETRIC MODEL

The research presented in this book spurred the development of a design and planning tool to determine optimal combinations of building and city improvements for highest carbon reduction and energy savings. This Parametric Model will function as an extension of the Data Model discussed elsewhere. Where the Data Model focuses on the existing condition of the city, this tool is concerned with the future of the city and the potential impact on performance and sustainability that discerning change will enable. As in the Data Model, the power of the Parametric Model is its potential to assess the city as a functioning whole, and to understand how local, bottom-up changes can have a great cumulative effect.

The buildings and other systems that make up the city in the Data Model are themselves collections of values and properties. For these calculations, a building has many values in addition to its geometry (such as gross floor area, age, number of inhabitants and distance from public transit) and many properties (type of occupancy, mechanical systems, facade and light fixtures). Thus, the buildings in the model are stored as collections of parameters, and the complex interaction of all these parameters comprises the city.

The Parametric Model allows for the manipulation of these parameters. The user makes adjustments to building parameters while the tool calculates the manifestation of these changes on the total carbon impact of individual buildings and the city as a whole. Designers, planners and building owners can take advantage of this tool to inform their decisions about upgrading mechanical, facade or lighting systems. The Parametric Model gives the user the ability to

test many different options, adjusting parameters in any possible combination. The tool also employs search heuristics to optimize the process of upgrading buildings so that the greatest savings in performance efficiency are executed with the given funds or investment.

The Parametric Model also incorporates citywide systems. For instance, the production method of the electricity consumed is critical to the carbon impact of each building; cleaner energy means smaller footprints. The balance of energy production methods is treated as a parameter, so the user can calculate the enormous affect that just a small percentage of wind energy can have on the footprint of the city. Also parameterized is transportation, another critical factor in the city's carbon footprint. Where the Data Model is designed to shed light on the performance of our urban environments, the Parametric Model is a crucial tool to enable their improvement.

User	Scale	Input	Model	Output
			Building Energy Analysis	
		How much funding do you want to allocate?	Construction Cost Estimate	Optimal funding allocation
Government Planner	Urban		Property Value Estimate	
		How much carbon do you want to save?	Carbon Impact Analysis	Optimal allocation and estimated cost
Building Owner/Manager	Campus		Urban Energy Analysis	
		What specific upgrades do you want to make?	Search Heuristic	Estimated cost and carbon reduction
Public	Building		Visualization Engines	
		What is the future of my city?		Projected and proposed plans

This diagram summarizes the overall Parametric Model. The tool is intended for multiple users with multiple intentions. The four modules under Input describe the four primary functions of the project.

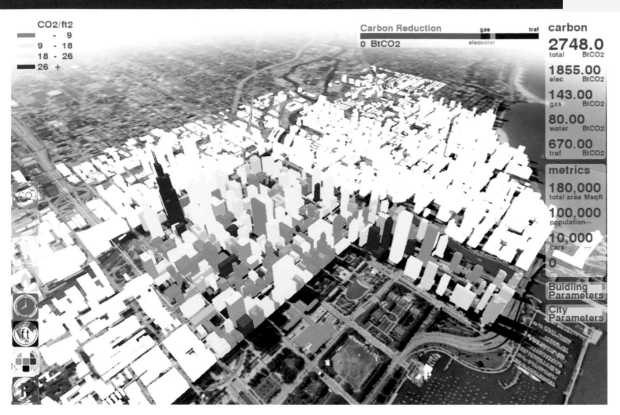

The basic interface is shown here with the buildings color-coded for total carbon use. This is a live 3D environment. The user is able to zoom, pan and orbit around all geometries. City totals for expended carbon are displayed at right, as well as other metrics such as total population, total square footage and parking totals.

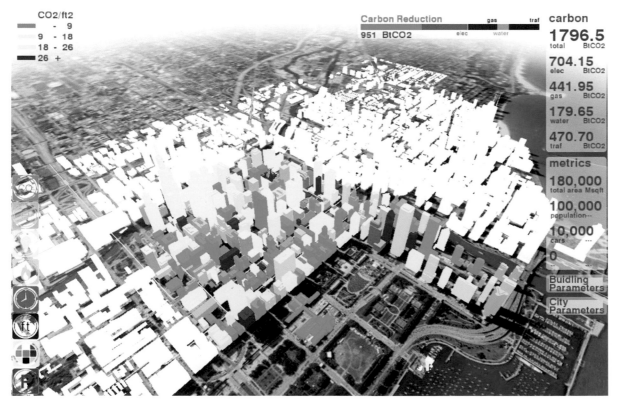

Here the global city parameter for ratio of commercial to residential space use has been adjusted. The resulting carbon reductions are registered in the color coding of certain buildings as well as in the overall totals and the Carbon Reduction bar in the upper right.

Every building in the study area can be accessed to understand its individual impact. The first level of basic data is shown here, including total carbon expended, as well as the contributions to the total by electricity, gas and water use.

Cultural
Gov. Office
Mercantile
Misc.
Office
Parking

Carbon Reduction   gas   traf
0 BtCO2   elec water

carbon
2748.0
total   BtCO2

1855.00
elec   BtCO2

143.00
gas   BtCO2

80.00
water   BtCO2

670.00
traf   BtCO2

AON CENTRE

carbon **41,759** MTCO2/yr

elec   **50,485,912** kWh/yr

gas   **1,097,390** therm/yr

water   **0** gal/yr

area   **3,243,640** ft2

use   **Office**

value   **0** $/yr

metrics
180,000
total area Msqft

100,000
population---

10,000
cars

0

Buidling Parameters

City Parameters

Retrofits can be tested for their impact on the total carbon use of the building, as well as their economic costs and benefits. Here the building-use ratio is adjusted, reflecting a fair amount of carbon savings.

Cultural
Gov. Office
Mercantile
Misc.
Office
Parking

Carbon Reduction   gas   traf
8 BtCO2   elec water

carbon
2740.0
total   BtCO2

1847.00
elec   BtCO2

143.00
gas   BtCO2

80.00
water   BtCO2

670.00
traf   BtCO2

AON CENTRE   RETROFIT

carbon **34,271** MTCO2/yr   reduction **7,488** MTCO2/yr

elec   **40,139,892** kWh/yr   investment **0** $ x1000

gas   **1,547,163** therm/yr   return period **0** yr

water   **0** gal/yr   operation savings **0** $/yr

area   **3,243,640** ft2

use   **Office**

Office   Residential
48%   51%
1,581,851 ft2   1,661,788 ft2

value   **0** $/yr   new value **0** $/yr

Residential/Office   0.48767793
0   ratio   1
0.0
0   1
0.5
0   1

metrics
180,000
total area Msqft

100,000
population---

10,000
cars

0

Buidling Parameters

Occupancy

Internal Demand

Mech Perf.

Envelope Perf.

Envelope Perf.

The built environment can be infinitely complex. The Parametric Model allows users to tame and efficiently manage mountains of data to enable informed, intelligent restructuring of cities toward a sustainable future.

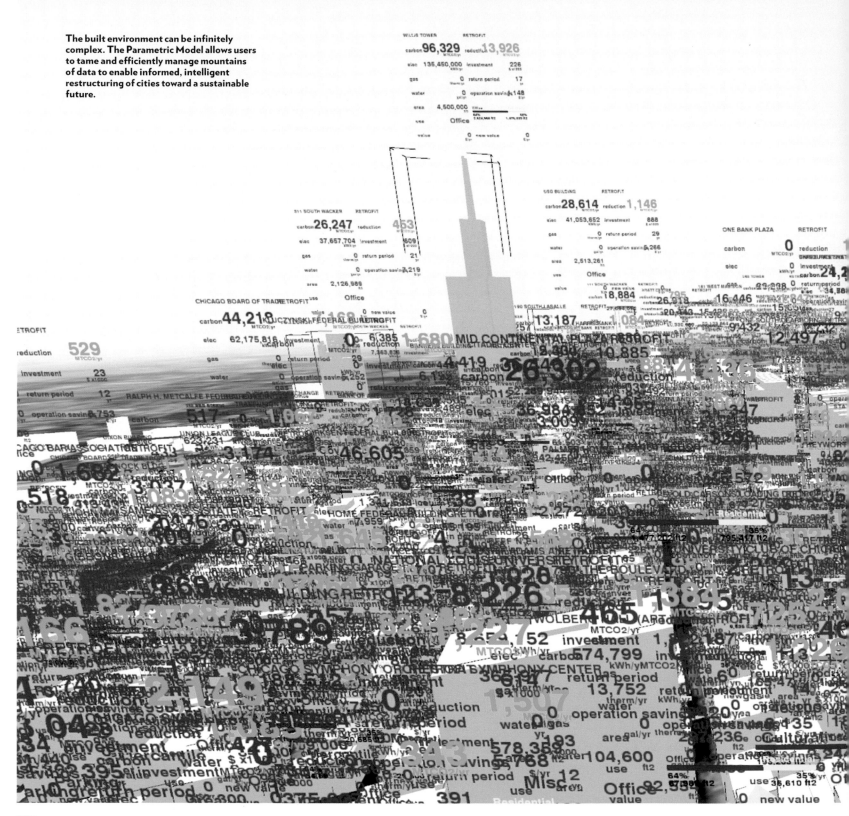

FUNDING

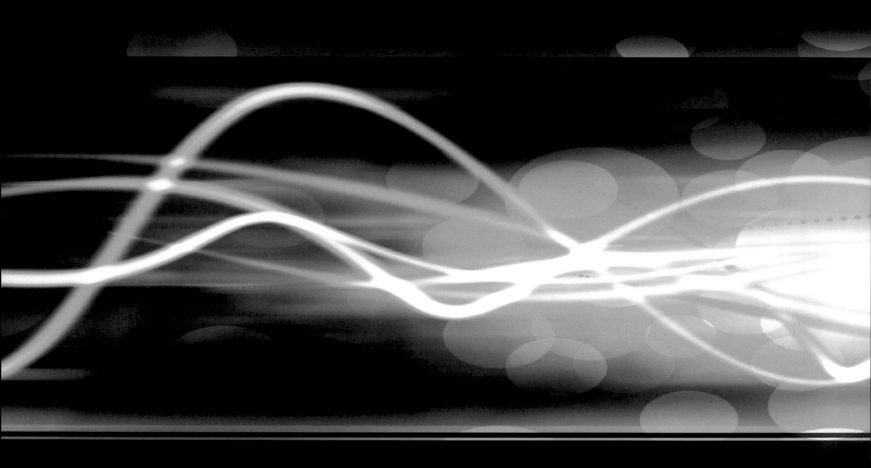

The DeCarbonization Plan has the potential to tap an array of funding resources at federal, state, municipal and private sector levels. These could work together to fund a successful DeCarbonization program in the Loop.

The Chicago Central Area DeCarbonization Plan has the potential to tap an array of funding resources. Several of these, at the federal, state, municipal and private sector levels, are described below. This chapter discusses these resources and examines which would work best for funding a successful DeCarbonization program in the Loop. We also identify a number of properties in the study area and discuss how these resources could be used to achieve DeCarbonization. A key consideration will be aligning the resources so they reinforce each other; so too with the properties. The overarching goal will be to integrate a number of the resources and gain the participation of several properties, so that a district is formed in which many owners undertake significant energy improvements. Two things are paramount: ensuring that major game changing DeCarbonization projects are launched, and creating a groundswell so that many DeCarbonization initiatives are undertaken by a wide group of owners in the wake of these pilot projects.

A partial list of a number of the federal programs:
- Energy Efficiency and Competitive Block Grants Program (EECBGs)
- State Energy Programs (SEP)
- Renewable Electricity Production Credits
- Energy Tax Credits
- Exempt Facilities Bonds (EFBs)
- Energy Efficient Commercial Building Property Deduction

These and other federal programs are described in the Appendix to this chapter.

A list of a number of the Illinois programs:
- State Energy Program
- Department of Commerce and Economic Opportunity (DCEO)

A partial list of Chicago's programs:
- Tax Increment Financing (TIF)
- Special Service Area Financing (SSA)

A partial list of Commonwealth Edison (ComEd) programs:

Prescriptive incentives
- Provides a menu of incentives for common commercial and industrial efficiency measures including lighting technologies, motors, HVAC and refrigeration (incentives are fixed and are paid on a per-unit basis)

Custom incentives
- Available for retrofit or major energy-saving measures not included in the list of qualified prescriptive measures
- Covers more complex measures including industrial improvements
- Incentives range from 3 cents to 7 cents per annual kWh saved
- Pre-approval required

Retro-commissioning
- Provides expert building analysis at no cost through a systematic evaluation of energy-using systems
- Can identify opportunities for customers to:
- Improve outside air control and indoor quality
- Eliminate "hot" or "cold" spots
- Optimize chiller or other equipment operation
- Reduce supply air temperature and fan speed in air handling units

New construction

- Provides new construction assistance and incentives to building designers, architects and owners for surpassing standard new construction building practices
- Two-track approach allows customers in various stages of design to participate in the program
- Buildings in early design stages can take advantage of whole building modeling and incentives
- Buildings that have completed their design will be able to qualify by specifying more efficient technologies to be implemented

Energy insights online

- Provides a free Web-based energy analysis service detailing your building's electricity consumption
- Users can download consumption data, including peak demand dates and times, weather temperature data and custom reports
- Accessible anywhere via the Internet

Any one of these programs, standing alone, is insufficient to accomplish the type of area-wide private investments in DeCarbonization improvements that are needed. Building owners and businesses that occupy these buildings repeatedly note that the decision to undertake substantial DeCarbonization improvements is a real challenge. There are a number of reasons for this. Debt and equity capital are needed to sustain the business enterprise itself. As a result, energy improvement investment is deferred or not even considered. Which measures will pay for themselves, at what cost, and over what period of time—these are the questions invariably raised by skeptical owners and their lenders.

Building owners and business operators do not generally undertake the life-cycle cost-benefit analysis necessary to understand the value of energy improvements. Therefore, any program we design must do several things:

- make clear the real benefit of energy improvements by enabling efficiency evaluations and commissioning services
- establish building-specific, non-weather-adjusted baselines, based on the 1990 CCAP target, against which energy savings can be measured
- identify a range of funding resources that will make feasible a building owner and/or a tenant business operator's investment in increased energy performance
- specify which funding resource is appropriate to the level of energy improvement being contemplated

This last point is crucial. Some owners are at the threshold of understanding. They need help comprehending the value proposition at a basic level. For them, efficiency evaluations and baseline commissioning activities need to be furnished. Funding resources for these costs could come from U.S. Department of Energy grants and Illinois's DCEO and ComEd programs. Indeed, the application for an EECBG grant from the DOE of up to $150 million provides the way to create a systematic approach to energy improvements in Chicago's Loop and its environs. Accordingly, we discuss next how a program could be established that uses (i) grants from the federal government and the State of Illinois, (ii) bonds issued by the City of Chicago using a variety of programs and funding sources and (iii) ComEd's resources, all in a systematic and integrated fashion.

The proposed program includes subsidized efficiency evaluations, direct subsidies and government-provided conduit financing related to energy efficiency improvements of commercial and public buildings throughout the Loop and its environs. The overall goal of this program would be to incentivize building owners to evaluate their properties and to make those energy-efficiency improvements that make the most sense for their specific properties. The types of energy-efficiency improvement projects likely to be included in such a program are:

· Re-commissioning of building systems
· Operator training
· Lighting fixture upgrades (upgrades to T8s, etc.)
· Installation of occupancy sensors
· Mechanical equipment upgrades (installation of VFDs, etc.)
· Plumbing fixture retrofits
· Any other measures for which an applicant could demonstrate meaningful energy efficiency improvement

The principal goal of this program is to reduce the carbon footprint of Chicago's Loop through the creation of a funding mechanism that provides sufficient incentives to cause building owners to implement energy-efficiency improvements. The amount of direct public subsidy required to achieve this result needs to be calibrated so that this financing program is far-reaching, sustainable and replicable. Accordingly, the proposed funding structure for these types of improvements uses a project finance approach in which the capital required for a particular improvement is a combination of both public and private debt and equity.

Public funding for the program will likely involve (i) outright subsidy of some portion of the design

and construction of improvements and (ii) conduit financing enabling building owners to pay the balance over an extended period at tax-exempt or otherwise subsidized interest rates.

The first step in this financing strategy requires some portion of each improvement (likely 5-15%) to take the form of a grant. A primary source of these grants could be the EECBG competitive grant funds soon to be offered by the U.S. Department of Energy (likely in grants of $50-150 million). These funds, which require at least a 5/1 non-federal/federal match, could be used to incentivize private building owners to make energy-efficiency improvements to their buildings. A first step will be the subsidy of efficiency evaluations conducted by pre-qualified firms, allowing building owners to appreciate the significant energy savings that properly designed improvements can create.

The second step involves the creation of public funding mechanisms available to help provide financing for energy-efficiency improvements. The key part of this second step is for the city to leverage its access to the public capital markets, thereby allowing a building owner to borrow funds on a conduit basis through the city at lower rates and repay this debt over an extended period. One obvious mechanism for such a financing vehicle is special service area financing. In special service area financing, a building owner agrees to submit to the formation of a special service area including its property (see Section 27-87 of the Illinois Special Service Area Tax Law). The city then levies a special tax against property within the special service area. This would allow the city to issue bonds secured by a special tax formulated according to a rate and method and then established in a special tax roll.

The proceeds of these bonds would be used to pay for a project's energy upgrades (or, ideally, would fund several different contiguous projects). The building owner pays the special tax on an annual basis, and the special tax is used to repay the bonds. In cooperation with engineering professionals, the building owner should be able to determine the estimated energy cost savings resulting from each such improvement, and to have the special tax levied at a rate so that the building owner (and/or tenants) pays no more in combined energy and special tax payments than they would have paid if the energy savings had not been generated.

Billing and collection of the special tax could occur in a number of different ways. One would be for the city to retain a fiscal agent who would bill and collect the special tax directly from property owners (similar to the city's past practice with special assessment district financings). Other alternatives worth exploring would be to (i) work with Cook County to have the special tax included on the county's annual tax bill; (ii) work with ComEd in to provide for the special tax to be billed on each user's energy bills or (iii) provide for the special tax to be paid on each user's water bills provided by the city. Each of these last three billing and collection possibilities has certain advantages over a fiscal agent billing and collection procedure. However, each would also require additional action by the county, the Illinois Commerce Commission or the city.

The city could establish an application procedure through which, by using the results of an energy audit and energy improvement cost estimates, the building owner could determine what energy efficiency improvements make the most sense. The owner could also see the shortfall, if any, between the cost of the improvement and the payback

(in energy savings) to be achieved as a result of performing the energy improvement project. With information about potential energy savings and improvement costs, the city could then determine, for a given applicant, (i) a level of direct subsidy for any given project and (ii) a level of conduit financing available for such a project. We strongly suspect that many energy improvements have not occurred because building owners have been unable to secure financing that would allow them to phase the costs over a reasonable payback period.

In summary, the steps to implement such a program include:

- Encouraging buildings to perform efficiency evaluations where such audits have not already been performed (with potential audit subsidies, if necessary).
- Creating an application process by which a building owner, using the results of an energy audit and cost estimates, would apply to the city for funding. Such an application would require the building owner, in the event that they are requesting direct subsidies, to demonstrate that the cost savings realized by such an improvement will not repay the cost of such improvement over a commercially reasonable period.
- Based on this application, the city would determine an appropriate level of public subsidy, potentially including both front-end grants to cover efficiency evaluations and public financing of the improvements using a combination of EECBG, QECGs, TIF bonds or notes and SSA bonds.
- The program should encourage building owners to use conduit public financing through an SSA mechanism to fund the proposed energy-efficiency improvements. The city would form an SSA, levy a special tax against the property within

the SSA, issue bonds secured by that special tax and use the SSA bond proceeds to pay for improvements. The SSA bonds would then be repaid either by a building owner directly paying a special tax or by separate billing and collection, by ComEd or the city through its water department, of charges (with the charges to be turned over to the city to pay the debt service on the SSA bonds) with the special tax lien acting as security to ensure that the charges are paid.

Example

This chapter has identified a range of possible energy efficiency improvements, many of which are applicable to a large number of buildings in our study area. For instance, an energy audit of a typical building may reveal many no-cost or low-cost improvements that can be made to enhance energy efficiency. However, in addition to these items, there are often capital-intensive items for which an owner will require some assistance. The program outlined above and on the previous page is intended to address these needs.

One example could be as follows. Building X has an energy audit performed. It determines that, in addition to several no-cost or low-cost solutions, substantial energy savings could be realized by replacing Building X's chiller with a new, more efficient one. Using an estimate for improvement arising out of such an audit, Building X is then able to determine that such a chiller replacement would cost approximately $4 million. However, Building X has several impediments to performing this upgrade on its own. First, to install a new 2,000-ton chiller, the capital needs of Building X's owner would normally include (i) ComEd incentives (which for a 2,000-ton chiller would be at most $80,000) and (ii) equity or a combination of equity and debt. However, in many situations, the projected cost savings from an upgraded chiller do not adequately repay the debt available from the traditional sources for such improvements, which in the current lending

environment are scarce. Therefore, Building X has two obstacles if it wants to upgrade its chiller: (i) a gap between a reasonable repayment period for its investment and (ii) a lack of access to financing.

Under the proposed program, Building X could approach the city and apply for benefits. To apply for a subsidy, Building X would have to demonstrate that the anticipated cost savings of the upgraded chiller does not repay the investment (in this case $3.92 million) over a reasonable period of time. If Building X could demonstrate such a shortfall, the city could provide an incentive, funded by EECBG funds, TIF funds or a combination of the two.

In addition, the city could also help to alleviate Building X's financing issues by establishing a special service area encompassing Building X and levying a special tax against Building X. The city can issue tax-exempt conduit SSA bonds at lower interest rates than Building X could obtain even if conventional financing were available, with the bonds secured by the special tax. The proceeds of these bonds would go to reimburse Building X for all or a portion of the $3.92 million investment in the new chiller. The bonds would be repaid over time by payment of the special tax levied against Building X.

Improvement type/potential funding service	QECB*	EECBG**	TIF	District Financing	ComEd***	DCEO***	Federal Loan Guarantee Program
Glazing upgrade	X	X	X	X			X****
Mechanical equipment upgrades (VFDs, others motors)	X	X	X	X	X	X	
Boiler plant upgrades	X	X	X	X			X****
Lighting fixture upgrades (Y8s, LEDs, etc.)	X	X	X	X	X	X	
Re-commissioning of building systems	X	X	X	X	limited	limited	
Chiller plant upgrades	X	X	X	X	X	X	X****
Occupancy sensors	X	X	X	X	X	X	

*Limited to capital improvements. Private sector investment limited to 30% of total funds unless given as a loan or grant in connection with a green community program.

**Energy Efficiency Block Grants—Approximately $27 million awarded to the City of Chicago to date. Allowable costs include energy audits and retrofits.

***Generally ComEd programs are targeted at industrial and commercial users while DCEO·s programs are targeted at public sector users. Both ComEd and DCEO programs for building re-commissioning are limited in the allowed number of participants.

****The Loan Guarantee Program is primarily concerned with improvements that "employ new or significantly improved technologies compared to technologies in service in the U.S. at the time the guarantee is issued."

Project launch proposal

This type of energy improvement program would best be launched in cooperation with a few identified properties. The goal would be to identify groups of buildings with owners or property managers who are committed to energy efficiency as well as the proposed program. These buildings could serve as demonstration projects for the program, showing how a combination of subsidies and financing can make energy-efficiency improvements possible. The city would approach such owners prior to the official program launch to get their commitments to the program, then combine the news of their participation with an official launch, at which time other building owners would be able to apply for benefits. This dual strategy is essential to show potential participants that such a program is viable and has the requisite private support.

A good list of candidates for the pilot projects, as well as large-scale districts, can be found in the buildings comprising the Chicago Green Office Building Challenge. We list these buildings by address and also note whether they are in existing TIF districts.

Building Location	Existing TIF District	Name of TIF District
11 S. LaSalle Street	YES	
10 S. Riverside Plaza	YES	
120 S. Riverside Plaza	YES	River West TIF
311 S. Wacker Drive	NO	LaSalle Central TIF
303 W. Madison Street	NO	
555 W. Madison Street	YES	
100 N. Riverside Plaza	YES	LaSalle Central TIF
980 N. Michigan Avenue	NO	
77 W. Jackson Blvd.	NO	
101 N. Wacker Drive	YES	
333 W. Wacker Drive	NO	
333 W. Wacker Drive	NO	
1 S. Dearborn Street	NO	
70 W. Madison Street	NO	
321 N. Clark Street	NO	
111 S. Wacker Drive	NO	
55 W. Monroe Street	YES	LaSalle Central TIF
225 W. Wacker Drive	NO	
550 W. Washington Street	YES	River West TIF
1 S. Wacker Drive	NO	
550 W. Adams Street	NO	
633 N. St. Clair Street	NO	
211 E. Chicago Avenue	NO	
550 W. Jackson Blvd	YES	Canal/Congress TIF
210 S. Canal Street	YES	Canal/Congress TIF
150 N. Michigan Avenue	NO	
71 S. Wacker Drive	NO	
135 S. LaSalle Street	YES	
200 E. Randolph Street	NO	
1 E. Wacker Drive	NO	
200 W. Monroe Street	YES	LaSalle Central TIF
181 W. Madison Street	YES	LaSalle Central TIF
50 W. Washington Street	NO	
222 Merchandise Mart Plaza	NO	
330 N. Wabash Street	NO	
330 N. Wabash Street #3600	NO	
1 N. Franklin Street	YES	LaSalle Central TIF
161 N. Clark Street	NO	
222 W. Adams Street	NO	
1 N. State Street	NO	
150 N. Wacker Drive	YES	LaSalle Central TIF
233 S. Wacker Drive	YES	LaSalle Central TIF
20 N. Michigan Avenue	NO	
737 N. Michigan Avenue	NO	

We further recommend the creation of the following pilot project areas:
Monroe/Madison buildings:
200 W. Monroe*
230 W. Monroe*
181 W. Madison*

LaSalle/Jackson buildings:
135 S. LaSalle—Bank of America*
231 S. LaSalle—Bank of America*
230 S. LaSalle—Federal Reserve*
141 W. Jackson—Chicago Board of Trade Building*

Lakeshore East buildings:
205 N. Michigan—Michigan Plaza South
225 N. Michigan—Michigan Plaza North
130 E. Randolph—Prudential Plaza 1
181 N. Stetson—Prudential Plaza 2
200 E. Randolph—Aon
233 N. Michigan—2 Illinois Center
111 E. Wacker—1 Illinois Center

*denotes properties located within LaSalle Central TIF

We've discussed the opportunities presented by the Loop's buildings at length because they represent the potential for the greatest energy savings (30%), and the part of the Loop ecosystem with the greatest number of financial tools (TIFs, SSAs, EECBGs and QECBs) most readily available to achieve those energy savings. But it's also important to discuss other factors that are major contributors to carbon production in the Loop. They, too, offer opportunities for carbon reduction, and the ways and means to finance those projects exist as well.

In the Urban Matrix chapter, we argue the merits of a 24/7 Loop with a strong mixed-use component that increases residential areas and provides a wider array of amenities. Existing tax increment financing districts can provide a powerful tool to foster this urban matrix. A number of these existing TIF districts (LaSalle Central, South Loop, Canal/Congress, River South and River West, for example) are contiguous. In their redevelopment plans, they have identified many opportunities to convert aging buildings to different uses. These plans also authorize the use of tax increment financing for modernization and renovation projects, which would support energy retrofits. They also authorize expenditures for public improvements that could take the form of pocket parks, permeable alleys and reduction in roadway widths to allow green connector passageways, all of which would reduce carbon.

So, too, with overlay special service areas. Apart from the renewals relating to mechanical and electrical systems we discussed in the Buildings portion of this chapter, special service areas could be used to finance green roofs for buildings. Each of those buildings could use a discrete special service area that could then be aggregated into a large district. The urban matrix could also be furthered through neighborhood schools in the Loop. Several

ways to help finance these schools exist under current redevelopment plans for a number of the TIFs in the Loop. Eligible redevelopment project costs include public improvements (new schools). The Illinois TIF Act also authorizes a set-aside of a portion of the tax increment arising from a residential project where any such residential project is given tax increment financing assistance.

Eligible redevelopment project costs also include those associated with vocational training. The opportunity for Loop businesses to provide training for high school and college students through tax increment grants covering the costs of these training activities also exists. Here the assistance from TIF could be simply enough to start such a program. It's likely that businesses in the Loop would embrace such a training opportunity as an instance of public service.

Daycare is another eligible redevelopment project cost. Here, too, a TIF subsidy for daycare could be small. Businesses will want to promote daycare that is convenient to attract and keep valuable employees. Special service area financing could provide additional support for businesses that want to increase benefits available to their employees as a way to enhance retention and also reduce the cost of vocational training, daycare and intermodal transport, including bicycling.

In addition to energy retrofits, tax increment and special service area financing could also be used to support adaptive reuses of the class B and C buildings described in this study. A number of the federal and state programs described in the Appendix to this chapter could be used in conjunction with the recommended municipal incentives.

The Mobility chapter presents the greatest opportunity for carbon reduction after the retrofit of existing buildings. The key to success here is interconnections. The goal should be an intermodal system that aids and abets all modes of transport—train, bus, car, bicycle and foot traffic. The Illinois TIF Law has a little-used provision that could enable a large intermodal TIF district. Such a district would be limited to intermodal costs such as station improvements, bicycle stands and pedways. Such an expanded district could provide some of the funding for the Carroll Avenue Transitway, the Clinton Avenue Subway and the East–West Transitway along Monroe, all of which are part of the Central Area Implementation Plan.

The key benefit in each of these major projects would not be the amount of the additional assistance. Rather, such an intermodal district would provide a means to share the local matching costs of such key projects across a broader local geography, giving those who benefit from such public improvements a way to participate, at least modestly, in their cost. The same logic could apply to a large overlay special service area created in tandem with such a large intermodal TIF transportation district. The creation of these two large districts could provide the impetus for much wider adoption of a number of pioneering projects such as bicycle paths and bicycle sharing, car sharing, green taxis and an expanded pedway system.

The opportunities presented by the Smart Infrastructure chapter are legion. Many of the programs described in the Appendix to this chapter are relevant. The EECBG grant from the U.S. Department of Energy could dovetail with the Smart Grid grant application being pursued by BOMA. A plug-in electric drive vehicle program could be considered. A special service area to enhance high-speed communication networks could also be considered for the Loop.

Saving water is another area in which a district approach makes sense. One good idea is linking appliance replacement with city water billing. The key is widespread buy-in. Systematic and broad-scale purchasing provides a means to save money on the cost of new water-saving plumbing fixtures. Such programs could be popularized through the strategic use of EECBG funds. Special service areas devoted to water-saving strategies and modernization projects could be established as part of the plan for the use of the EECBG grants. The special services for such districts could support toilet fixture rebates and advertising programs as well as large-scale greywater systems.

Waste disposal and transport provides a similar opportunity. The coal tunnel system underneath the Loop could be outfitted with pneumatic tubing. A special service area could equitably distribute the cost of such a system, or it could simply be the subject of a solid waste disposal system operated on an area basis by a private concern, which would charge a fee for collecting and pneumatically transporting the waste to a plasma arc furnace near the Loop. Chicago has the benefit of abandoned industrial areas near its central business district. Indeed, the pneumatic system of disposal and the plasma arc furnace could both be the subject of a creative project finance transaction that would use the feedstock of waste to generate the source of vaporizable material. The plasma arc furnace separates the waste stream into three components: (i) vapor for much of the waste (hence no emissions), (ii) a glassy aggregate (that is marketable as a construction fill substrate) and (iii) molten metals (that also have an independent market value). This way, feedstock becomes byproducts that provide revenue and lower the cost of the waste transport and disposal.

Conclusion

This discussion demonstrates the manifold ways in which the DeCarbonization Plan's carbon reduction strategies work together. The financing resources and strategies discussed can also work together, strengthening each component part and making the achievement of carbon reduction goals more likely.

FUNDING CHAPTER APPENDIX

Program	Overall Goal	Funding Opportunities	Links / Applications
Energy Efficiency and Conservation Block Grant Program (EECBG)	The EECBG program is intended in part to assist cities in their efforts to improve energy efficiency in the transportation and building sectors. Grants can be used for energy efficiency and conservation programs. According to the EECBG website, this includes but is not limited to: (1) building energy audits and retrofits, including weatherization; (2) financial incentive programs for energy efficiency; (3) building code development, implementation, and inspections; (4) installation of distributed energy technologies including heat and power and district heating and cooling systems; and (5) installation of renewable energy technologies on government buildings. This could be a potential source of funding.	ARRA includes $3.2 billion in funding for the EECBG program. $2.7 billion of this will be awarded through formula grants. More than $400 million will be allocated through competitive grants, which will be awarded through a separate future Funding Opportunity Announcement (FOA). **Illinois Department of Commerce and Economic Opportunity (DCEO)—$21 million** Under the EECBG program, the state of Illinois will receive $112,175,600. The DCEO will receive $21,834,600. Of this, 60% must be used to provide grants to cities that are not eligible for direct grants from the Department of Energy. The remaining 40% will go to projects identified by the Bureau of Energy and Recycling (a subdivision within the DCEO). According to the DCEO website, the types of projects likely to be covered may include buildings. **City of Chicago—$27 million** Additionally, the City of Chicago received $27,648,800. The City of Chicago currently estimates that it will distribute these funds in the following way: • $8,000,000 will go to the Department of General Services (DGS) for "City Facility Energy Improvements." • $6,500,000 will go to the Department of Street and Sanitations, Bureau of Electricity (DSS-BOE) for "Streetlight Efficiency Upgrades."	EECBG Information DCEO Website Chicago Budget for EECBG Funds

		• $750,000 will go to the DGS for "City Facility Parking Lot lighting Upgrades." • $5,099,000 will go to the Department of Energy and Department of Application Community Development DOEDCD) for "Residential Energy Efficiency." • $7,300,000 will go to the DSS-BOE for "LED Traffic lights at City/State Owned Intersections." **Application deadline:** The application deadline was May 26, 2009 for states and June 25, 2009 for local governments. Both Illinois and the City of Chicago applied for and were awarded grants.	Application
Weatherization Assistance Program	The goal of WAP is to reduce energy costs for low-income families by improving the energy efficiency of their homes. It's unlikely that this program would be a good source of funding as it's used exclusively to fund residential buildings.	The ARRA includes $5 billion for the Weatherization Assistance Program. The State of Illinois has been awarded $242,526,619. This funding is available from April 1, 2009 through March 31, 2012. Approximately 40% of the ARRA funds will be targeted towards the 2010 Illinois Home Weather Assistance Program (IHWAP). The remaining 60% will be distributed in 2011.	Program Information
State Energy Program	The DOE's State Energy Program provides grants to the states to carry out their own renewable energy and energy efficiency programs. This could be a potential source of funding.	The ARRA includes $3.1 billion for SEP. Formula grants from the SEP will go to the Illinois DCEO. Illinois formula allocation for SEP under the ARRA is $101,321,000. The DCEO's website broadly states that the types of projects likely to be covered may include "renewable energy" and energy efficiency." The DCEO has yet to provide an application on its website as of this writing. **Application deadline:** Comprehensive application was due May 12, 2009. The DCEO applied.	Illinois DCEO Website Examples of how Illinois has used the money

Transportation and Electrification Grant Program	The DOE is seeking applications for grants that encourage the use of plug-in electric drive vehicles or that implement electric transportation technologies that would significantly reduce greenhouse gas emissions. It's unlikely that this program would be a good source of funding.	The ARRA includes $400 million for this program. Applications were due May 13, 2009.	Program Information
Clean Cities: Alternative Fuels and Advanced Technology Vehicles Pilot Program	This four-year program includes two years of data collection and two years in which projects must be completed. Up to 30 grants, based on geography, were awarded in fiscal years 2009 and 2010, ranging from $5 million to $15 million each. There is a 50% matching requirement. Eligible projects are those that further the domestic manufacture and use of alternative fuels. Eligible entities are limited to state or local governments or metropolitan transportation authorities, or any combination of these, as long as they are in partnership with a designated Clean Cities Coalition.	The ARRA includes $300 million for the DOE's Clean Cities Program. The first round of applications was due on May 29, 2009. The second round of applications was due September 30, 2009. The $300 million is only available until Sept. 30, 2011.	Program Information
Loan Guarantee Program	The DOE may issue loan guarantees to eligible projects that "avoid, reduce, or sequester air pollutants or anthropogenic emissions of greenhouse gases" and "employ new or significantly improved technologies as compared to technologies in service in the U.S. at the time the guarantee is issued." The DOE has broad authority to guarantee loans that support early commercial use of advanced technologies, if "there is reasonable prospect of repayment of the principal and interest on the obligation of the borrower." This is targeted at early commercial use only. In the first round, the DOE evaluated loan guarantees for projects that employed technologies in hydrogen, solar, wind and hydropower.	The ARRA includes $6 billion for the cost of guaranteed loans. Despite Secretary Steven Chu's announcement in February 2009 that the loan process would be expedited, and to the disappointment of the alternative energy lobby, new application guidelines are still unavailable as of this writing. **Application deadline:** DOE has not made application guidelines available.	DOE Link The application is not posted online.

Section	Title/Description/Qualification Specifics	Applicability to Sustainable Energy Initiatives	Limitation	Length
26 USCS 45	Renewable Electricity Production Credit: Provides renewable energy production credits. Credit per taxable year equal to: 1.5 cents multiplied by the kWh of electricity produced by taxpayer from qualified energy sources at a qualified facility during a 10 year period beginning on date facility originally placed in service and sold by the taxpayer to an unrelated person during the year. (This 1.5 cents is not current; instead, the levels are 2.1 cents per kWh for wind, solar and geothermal and 1 cent for other qualified facilities.) "Qualified Energy Sources": Wind, closed-loop biomass, open-loop biomass, geothermal energy, solar energy, small irrigation power, municipal solid waste, qualified hydropower production and marine and hydrokinetic renewable energy. "Qualified Facilities": Facilities that produce qualified energy sources (each has its own specific definition). "Wind Facility": Any facility using wind to produce electricity owned by TP that was put in place by TP after December 31, 1993 and before Jan. 1 ,2010; this does not include any facility with respect to which a small wind energy property expenditure (Section 2sD (d)(4)) is taken.	The ARRA provides a three year extension on the production tax credit (PTC), while offering expansions on and alternatives for tax credits on renewable energy. The extension keeps the wind energy PTC in effect through 2010. PTC: The PTC provides a credit for every kWh produced at new qualified facilities during the first 10 years of operation, provided the facilities are placed in service before the tax credit's expiration date. For 2008, wind, solar and geothermal facilities earned 2.1 cents per kWh , while other qualified facilities earned 1 cent per kWh. Investment credit: As a result of the current slump in business activity and the fact that fewer businesses are seeking tax credits, the ARRA allows owners of nonsolar energy facilities to make an irrevocable election to earn a 30% investment credit rather than the PTC. Grant: Alternatively, the facility owner could choose to receive a grant equal to 30% of the tax basis for the facility. The grants are also available for renewable energy facilities that would normally earn a business energy credit 10-30%, including systems using fuel cells, solar, energy, small wind turbines, geothermal energy, microturbines and combined heat and power (CHP) technologies. To earn a grant the construction must be in either 2009 or 2010 and must be completed prior to the termination of the	Credit is marginally reduced when the reference price (annual average contract price per kW) is between 8 and 10 cents, and eliminated completely when the reference price reaches 11 cents (i.e., credit is reduced by a ratio equal to [the amount by which the reference price exceeds eight cents] divided by [three cents]).	A 10-year period beginning on the date the facility was originally placed in service; facility must be placed in service prior to January 1, 2010.

		PTC.
		Applicability to private sector: It's possible that the PTC, investment credit or grant program may apply to wind turbines and/or solar panels installed on buildings. Whether this credit will be of use to the project will depend on gathering more facts and likely being able to structure any energy-producing assets (solar, wind, etc.) into separate ownership.
		Overview of Tax Credits
		ARRA Text (Relevant pages are 33-39)

26 USCS 48	Calculation of the energy credit (part of the investment tax credit in Section 46). Energy credit is the **energy percentage** of the basis of each energy property placed in service during such taxable year.

"Energy Percentage": 30% for qualified fuel cell, energy and small wind energy property; 10% for other energy property.

"Energy Property": Equipment using solar power to generate electricity, to produce or distribute energy from a geothermal deposit, qualified fuel cell or microturbine property, qualified small wind energy property, etc.

"Qualified small wind energy property": Property that uses a turbine with a capacity of no more than 100 kW.

"Qualified fuel cell property": A fuel cell power plant that has a nameplate capacity of at least .5 kWh of electricity using an electrochemical process, and has an electricity-only generation efficiency greater than 30%. | Applicability to private building owners: This could apply to wind turbines or other alterative energy applications installed on site. For wind turbines, the nameplate capacity cannot be more than 100 kWh, allowing for a 10% energy credit.

This may apply to the cogeneration unit project allowing for a 30% energy credit. Note that any facility receiving a tax credit under Section 45 is not eligible under this program (see last line of Section 45(a)(3). | | Expires January 1, 2017 |
| 26 USCS 54 | Credit to holder of clean energy renewable bonds: Tax credit issued to holders of clean energy renewable bonds. Clean energy renewable bonds are bonds issued to fund any qualified project. | Applicability to private building owners: Clean Energy Renewable Bonds (CREBs) are generally intended to finance public power systems and cooperative energy producing entities. Therefore this | Credit may not exceed the excess of (the sum of regular tax liability plus tax imposed | Bonds must be issued before December 31, 2009. |

	"Qualified Project" is defined as any "Quality facility" as defined in Section 45 above. Annual tax credit equal to credit rate (determined by the Secretary) times the outstanding face value of the bond. The credit amount is determined on each of four credit increment dates each year; the amount earned on each date is equal to 25% of the annual credit amount.	program will likely not be applicable to the current project.	under section 55) over (the sum of credits allowed under Part IV of the tax code, with the exception of sections 31-37 and section 54A - 54D). Note: Sections 142(1) and sections 179(D) are not part of Part IV.
26 USCS 54D	Qualified energy conservation bonds: Creates a new category of bond issued by state and local governments for the purpose of funding one or more qualified conservation purposes. "Qualified Conservation Purpose"; Includes wind facilities as defined by Section 45 above, along with: • Capital expenditures for the purposes of reducing energy consumption in public buildings, implementing green community programs, etc. • Mass commuting facilities • Various research expenditures • Demonstration projects designed to promote the commercialization of green building technology, conversion of agricultural waste for use in the production of fuel, advanced battery manufacturing technologies, technologies to reduce peak use of electricity, technologies for the capture of CO_2 emitted from combusting fossil fuels to produce electricity and public education campaigns to promote energy efficiency.	Applicability to private building owners: This may be applicable to provide financing. Chicago's allocation is only about $32 million. Proceeds must go toward the payment of capital improvements. Only 30% may go toward private uses. One exception to this rule is if the funds are used for grants and/or loans to private owners in connection with a "green community program." The term "green community program" is not defined but some commentary suggests that this could include building greening and retrofit programs.	100% of the bond must be used for qualified conservation purposes. National qualified energy bond is limited to $3.2 billion. This is allocated to states based on population, then allocated to large local governments (100,000 persons) based on the ratio of local government population to the population of the state.	 
26 USCS 142 (I)	Exempt facility bonds: qualified green building and sustainable design projects. Any bond issued where 95% or more of the proceeds are to be used to provide, among others, qualified green building or sustainable design projects.	"Qualified Green Building Project": Any project the Secretary (after consultation with the EPA) designates as a green building and sustainable design project that meets the following requirements:	Project must be nominated by a state or local government within 180 days of the enactment of the subsection and state or local governments	

| | | Project proposal must describe energy efficiency, renewable energy and sustainable design features, and demonstrate that the following criteria are satisfied:
(i) At least 75% of the square footage of the building must be registered for LEED certification and is reasonably expected to receive such designation.
"Qualified Conservation Purpose": Includes wind facilities, along with:
• Capital expenditures for the purposes of reducing energy consumption in public buildings, implementing green community programs, etc.
• Mass commuting facilities
• Various research expenditures
• Demonstration projects designed to promote the commercialization of green building technology, conversion of agricultural waste for use in the production of fuel, advanced battery manufacturing technologies, technologies to reduce peak use of electricity, technologies for capturing CO2 emitted from combusting fossil fuels to produce electricity and public education campaigns to promote energy efficiency.
(ii) The project must include a brownfield site as defined by CERCLA.
(iii) The project must receive state or local support in an amount of at least $5 million (including tax abatements and contributions in kind).
(iv) The project must include at least 1 million square feet of building or at least 20 acres. The project must also provide permanent employment for at least 1,500 full-time equivalents when completed and 1,000 full-time equivalents during construction.
Applicability to private building owners: The brownfield requirement is a large hurdle but, if a given site meets it, this could be a powerful tool for buildings wishing to include "demonstration" type projects. | must certify in writing that the project will meet the requirements.

Total assurance of green building bonds may not exceed $2 billion. | |

251

| 26 USCS 179D | Energy efficient commercial buildings deduction: Deduction for the cost of energy efficient commercial building property placed in service during the taxable year. | "Energy Efficient Commercial Building Property": Property (a) with respect to which depreciation (or amortization) is allowable; (b) installed in a building located in the United States and within the scope of Standard 90.1-2001 of American Society of Heating, Refrigerating, and Air Conditioning Engineers and the Illuminating Engineering Society of North America; and (c) installed as part of the interior lighting systems, heating, cooling, ventilation and hot water systems, or as part of the building envelope.

The installed interior lighting, heating, cooling, ventilation, hot water systems and building envelope must either reduce the total annual energy and power costs by 50% or more, or be part of a taxpayer's overall plan, that includes the subsequent installation of the same, and will reduce the total annual energy and power costs by 50% or more, as compared with a "Reference Building," which is compliant with the minimum requirements of Standard 90.1-2001.

Applicability to private building owners: May allow owner to deduct for a portion of the costs that are not otherwise funded. | Maximum deduction is equal to $1.80 per square foot, less the aggregate amount of the Section 179D deductions allowed with respect to the building for all prior taxable years.

If the qualifications are only partially met, the maximum deduction is equal to $0.60 per square foot. | 
 |

Small wind turbines

- Renewable Electricity Production Tax Credit (PTC)
- Business Energy Investment Tax Credit (ITC)
- Renewable Energy Grants (authorized by ARRA) allow a grant in lieu of Business Energy Investment Tax Credit
- Federal Loan Guarantee Program
- Renewable Energy Production Incentive (REPI) (credit to public generators of electricity)
- Energy Efficiency and Conservation Block Grant Program (EECBG)

Photovoltaic (traditional and concentrated)

- Business Energy Investment Tax Credit (ITC)
- Renewable Electricity Production Tax Credit (PTC)
- Renewable Energy Grants (authorized by ARRA) allow a grant in lieu of Business Energy Investment Tax Credit
- Federal Loan Guarantee Program
- Renewable Energy Production Incentive (REPI) (credit to public generators of electricity)
- Energy Efficiency and Conservation Block Grant Program (EECBG)

Solar thermal

- Residential Renewable Energy Tax Credit
- Business Energy Investment Tax Credit (ITC)
- Renewable Electricity Production Tax Credit (PTC)
- Renewable Energy Grants (authorized by ARRA) allow a grant in lieu of Business Energy Investment Tax Credit
- Federal Loan Guarantee Program
- Energy Efficiency and Conservation Block Grant Program (EECBG)

Microturbines

- Business Energy Investment Tax Credit (ITC)
- Renewable Energy Grants (authorized by ARRA) allow a grant in lieu of Business Energy Investment Tax Credit
- Federal Loan Guarantee Program
- Energy Efficiency and Conservation Block Grant Program (EECBG)

Fuel cells

- Business Energy Investment Tax Credit (ITC)
- Renewable Energy Grants (authorized by ARRA) allow a grant in lieu of Business Energy Investment Tax Credit
- Federal Loan Guarantee Program
- Energy Efficiency and Conservation Block Grant Program (EECBG)

Biogas

- Business Energy Investment Tax Credit (ITC)
- Renewable Electricity Production Tax Credit (PTC)
- Renewable Energy Grants (authorized by ARRA) allow a grant in lieu of Business Energy Investment Tax Credit
- Renewable Energy Production Incentive (REPI) (credit to public generators of electricity)
- Energy Efficiency and Conservation Block Grant Program (EECBG)

Biofuel

- Business Energy Investment Tax Credit (ITC)
- Renewable Energy Grants (authorized by ARRA) allow a grant in lieu of Business Energy Investment Tax Credit
- Federal Loan Guarantee Program
- Energy Efficiency and Conservation Block Grant Program (EECBG)

Geothermal

- Business Energy Investment Tax Credit (ITC)
- Renewable Electricity Production Tax Credit (PTC)
- Renewable Energy Grants (authorized by ARRA) allow a grant in lieu of Business Energy Investment Tax Credit
- Federal Loan Guarantee Program
- Renewable Energy Production Incentive (REPI) (credit to public generators of electricity)
- Energy Efficiency and Conservation Block Grant Program (EECBG)

IMPLEMENTAT
STRATEGY

Ogilvie
Station

Clinton

Monroe

Intermodal
Station

Adams

Quincy

Union
Station

Clinton

Canal

Wacker

Franklin

Wells

ON

Clark

Dearborn

State

Wabash

Michigan

Grant Park

Implementing the Chicago Central Area
DeCarbonization Plan will take a concerted
effort that encompasses all eight of the plan's key
strategies.

The Chicago Central Area DeCarbonization Plan is organized around eight strategies for carbon reduction. The Monroe Corridor Pilot touches all eight of these strategies and sets up synergistic relationships that create greater carbon savings and lifestyle enhancements than if they were implemented separately. The Monroe Corridor is defined by the blocks and buildings adjacent to Monroe on the north and south. The east boundary is the lakefront; the west boundary is Clinton. Buildings along this corridor account for 15.6% of the square footage of the Loop's 184 million occupiable square feet.

This pilot project can be seen as a catalyst for further city development. It was selected based on four criteria:

1. Feasibility: The pilot project should be completed within five years.
2. Relevance : The pilot project should not conflict with existing city goals and plans such as the Chicago Central Area Action Plan (CCAAP).
3. Visibility: The pilot project should have a verifiable outcome and demonstrate change in the community or study area.
4. Leadership: The pilot project should have a sponsor or be led by an agency with the capacity to implement it.

STUDY AREA

Clinton　Canal　Wacker　Franklin　Wells　LaSalle　Clark　Dearborn　State　Wabash　Michigan

Monroe

The Loop currently consists of 184 million square feet of occupiable space. The Monroe Corridor is approximately 28.7 million square feet, or 15.6% of the total. The current emissions of the Monroe Corridor are potentially .6-8 MMTCO2e.

BUILDINGS AND ENERGY: RETROFIT STRATEGIES

To meet 80% targeted reductions in half of the Monroe Street buildings, a series of steps is needed.

- Recommendation guidelines should be issued for low-cost sustainability measures for all Loop study area buildings.
- For Monroe Corridor buildings, further levels of retrofit design and ESCO funding support should be identified.
- For targeted Monroe Corridor buildings, design, bid, build, commissioning, metering, and publishing of findings should be undertaken.

**Existing Energy Use of Monroe Corridor Buildings:
Red = Highest Energy Use, Green = Lowest Energy Use**

**Proposed Energy Use of Monroe Corridor Buildings:
Red = Highest Energy Use, Green = Lowest Energy Use**

CHICAGO CODE REVIEW FOR DECARBONIZATION

To achieve the goals of the Chicago Climate Action Plan, it's critical for the city to conduct a full review and re-write of the current zoning and building codes. Specific constraints for newly constructed and renovated buildings, both residential and commercial, need to be mandated.

At a minimum, a full code update should mandate:

· 20% water reduction for flow and flush fixtures
· 75% construction waste diversion
· Use of low-pollutant paints, carpets and floorings
· Commissioning of mechanical equipment
· Installation of water meters
· Public access to energy use data
· 20-30% over ASHRAE energy reduction
· Minimum LEED Silver for multi-family and commercial buildings
· Night sky requirements

© iStock/Thaddeus Robertson

CITY-WITHIN-A CITY-CONCEPT

Several existing buildings on Monroe Street are Class C office space. They have a high vacancy rate that could be converted for residential use, thus providing housing for new downtown residents.

There are other buildings on Monroe that are low-density, non-landmark structures that could be razed to provide sites for new high-performing, high-density office and/or mixed-use spaces. These new buildings are needed to meet the expected growth projected by the Plan.

A new school is required to accommodate these new residents and attract them to the Loop's urban lifestyle. New residential amenities, such as grocery stores, services and outdoor spaces (pocket parks and occupiable green roofs, for example) are also needed.

1 **Amalgamated Bank building, currently low-density, low-performance**

2 **Bank of America/Walgreens, currently low-performing, under-occupied**

3 **122 W. Monroe, on the northeast corner of Monroe and LaSalle**

NECESSARY STEPS

- Residential conversion
- High-performance, high-density building replacement study
- Add public schools and daycare centers
- Expand residential amenities and services—grocery stores, farmers' markets, etc.
- Biodiversity, pocket parks, occupiable green roofs and vegetated walls

MONROE CORRIDOR BELOW GRADE

The Chicago Central Area Action Plan envisions a below-grade East–West Transitway running from the West Loop train terminals to Grant Park, with an intermodal or West Loop Transportation Center located beneath Clinton. There is currently no convenient method of mobility from the west side of the Loop to the east side. The below-grade line could also connect to existing and new amenities to support the increase in residential population by incorporating an improved and extended pedway system.

KEY ELEMENTS
• Intermodal station
• East–West intermodal link; mode to be researched
• Connection to above grade
• Below-grade amenities: health clubs, daycare, grocery stores
• Improvements to existing pedway system
• New links to existing pedway system
• Links to existing transit—for example, a link to Union and Ogilvie Stations, Blue Line, Red Line, O'Hare

Central Station, Berlin

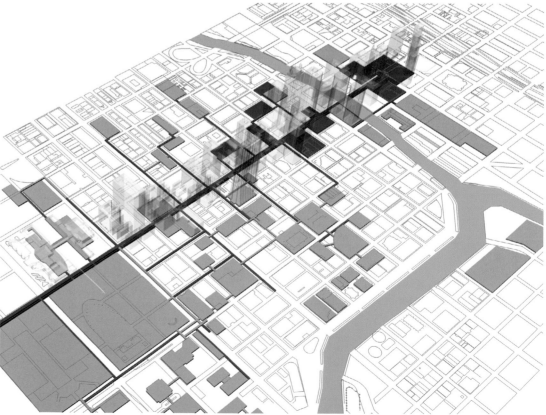

MONROE CORRIDOR ABOVE GRADE

KEY ELEMENTS
• Reduce Monroe Street to two lanes
• Increased vegetation along street edges
• Bicycle and pedestrian paths
• Bicycle stations
• One-way car-share station

A dedicated below-grade transit line will decrease traffic on Monroe, allowing for an improved bicycle and pedestrian experience above grade. Greening of this strip will provide a public space for outdoor events, such as markets and festivals, and reduce the urban heat island effect. Opportunities for mobility mode changes, such as a one-way car sharing depot and a west-end bicycle station, will be incorporated into the intermodal station at grade level. Improvements to the river edge bicycle/pedestrian path are also needed.

Las Ramblas, Barcelona

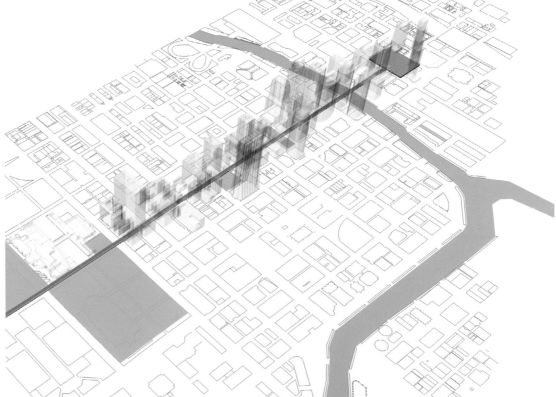

SMART GRID AND NET METERING

Implementing the DeCarbonization Plan depends in part on the design and deployment of a framework for resource monitoring and control of energy, water, waste water, lighting levels, indoor air quality, stormwater and other elements. This framework would also include strategies for resource-sharing among properties, including thermal, electrical and computing resources.

LIVING MACHINE DEMONSTRATION PROJECT

As a demonstration project only, a Living Machine could be incorporated into a new public environmental school to teach the sustainable and scientific principles involved in the use, treatment and re-use of water.

© Wikipedia/Conor Lee

FLUSH AND FLOW FIXTURE REBATES

A commercial and residential rebate program for retrofitting sanitary fittings could be launched to raise public awareness of the rebate program as well as the Chicago Climate Action Plan. The result would be savings in water consumption, water/waste water pumping costs and waste water treatment. An extensive advertising campaign to garner support and raise public awareness would draw attention to the carbon emissions savings associated with lower water consumption.

© iStock/Allsee

PNEUMATIC WASTE COLLECTION SYSTEM

A vacuum waste collection system will be used to collect solid recyclable and non-recyclable waste from residential and commercial properties. The system will result in increased recovery rates for recyclable waste (and therefore less waste to the landfill, resulting in lower greenhouse gas emissions), improved hygiene levels, reduced vermin and pests within city alleyways, and lower carbon emissions associated with waste collection via garbage trucks.

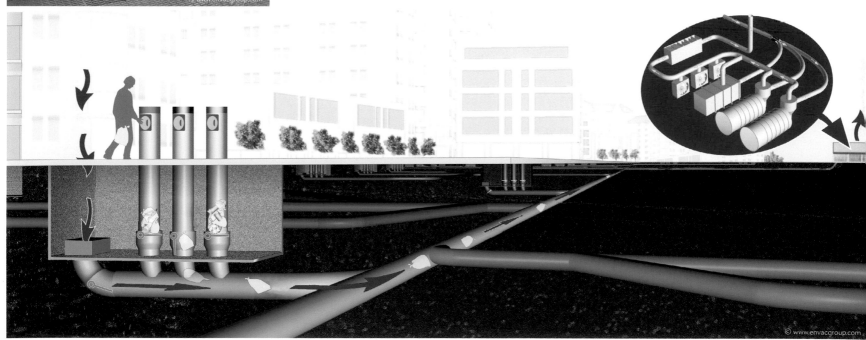

The Monroe Corridor and potentially the existing underground tunnel system could be used to pilot an automated solid waste collection system.

THE GREEN CITY

A new textbook called *The Green City* will be written as a summary of Chicago's vision for a carbon-free future and taught as part of the curriculum at Chicago public schools. The content will explain the human and environmental benefits of the Plan, as well as introducing the concept of the responsibility of residents as stewards of a great city. Small individual contributions and decisions, the book will emphasize, ultimately have the greatest impact on a city's sustainable future.

URBAN AGRICULTURE

The City of Chicago can play a key role in modeling for Loop residents how they might grow some of their own food. There could be an organic urban agriculture demonstration project in Grant Park, with the resulting produce sold at marketplaces throughout the Loop. This would serve as an educational example of the potential for urban agriculture in the city.

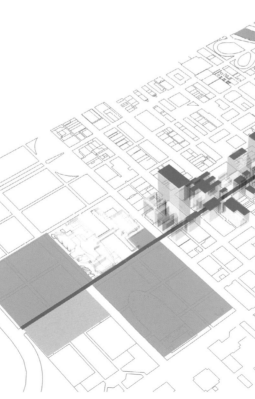

FURTHER DESIGN AND DEVELOPMENT OF MODELING TOOLS

The Data Model focuses on the existing condition of the city and can be used for data collection and benchmarking of both specific buildings and the city as a whole. The Parametric Model's function is to allow us to create and analyze retrofit and renewable strategies, and to provide a final carbon reduction total.

CO2/ft2
-   -   9
- 9 - 18
- 18 - 26
- 26 +

Carbon Reduction   gas   traf
5 BtCO2   elec water

carbon

2743.0
total   BtCO2

1850.00
elec   BtCO2

143.00
gas   BtCO2

80.00
water   BtCO2

670.00
traf   BtCO2

metrics

180,000
total area Msqft

100,000
population--

10,000
cars   ---

0
..   ---

AON CENTRE

carbon **37,688**
MTCO2/yr

elec   **44,861,408**
kWh/yr

gas   **1,341,904**
therm/yr

water   **0**
gal/yr

area   **3,243,640**
ft2

use   **Office**

value   **0**
$/yr

RETROFIT

reduction **4,071**
MTCO2/yr

investment   **0**
$ x1000

return period   **0**
yr

operation savings   **0**
$/yr

Office
72%   27%
2,340,226 ft2   903,413 ft2

new value   **0**
$/yr

Buidling Parameters

Occupancy

Internal Demand

Mech Perf.

Envelope Perf.

Envelope Perf.

MONROE CORRIDOR

The downtown Loop is currently comprised of 184 million square feet of occupiable space. The Monroe Corridor is about 28.7 million square feet, or 15.6% of the total Loop area. The current emissions of the Monroe Corridor are potentially .6-8 MMTCO2e. Eliminating those emissions will go a long way toward realizing the carbon reduction goals of the 2030 Challenge.

Current Emissions for Monroe Corridor (in MMTCO2e)	**.608**
Reductions from:	
Buildings and Energy	.247
Urban Matrix	.084
Mobility	.023
Smart Infrastructure	.016
Water	.002
Waste	.012
Community Engagement	.014
Total Reduction	**.398**
Remaining emissions from further renewable energy	.210
Goal in 2030	**0**

THE CHICAGO CENTRAL AREA DECARBONIZATION PLAN PROJECT TEAM

Senior Team

Adrian Smith, Partner
Gordon Gill, Partner
Robert Forest, Partner
Sara Beardsley, Senior Architect
Gail Borthwick, Senior Architect
Jeffrey Boyer, Senior Designer/Engineer
Carrie Neill, Director of Communications
Kevin Nance, Director of Public Affairs

Team

Miguel Alvarez, Rachael Bennett, Nathan Bowman, Hayes Brister, Gabriel Bunea, Ashley Byers, Christopher Drew, Michelle Dumont, Iris Gan, Gregg Herman, Brian Hubbard, Christopher Hurst, Brian Jack, Peter J. Kindel, Jocelyn Moriarty, Tyler Noblin, Matthew Nyweide, Jonathan Orlove, Luis Palacio, Robert Perry, Silviu Petrea, David Rariden, Dennis Rehill, Sean Satterfield, David Schweim, Daniel Segraves, Jason Smith, Katherine Smith, Jeffrey Stafford, Leslie Ventsch, Michael Waldo, Brent Watanabe, Brad Wilkins, Marc Woodcock

Consultants/Partners

Greg Hummel and Karl Marschel, Bryan Cave LLP
Commonwealth Edison
Edna Lorenz, Environmental Systems Design, Inc.
Peoples Gas
Commissioner Suzanne Malec-McKenna and the Mayor's Green Team, City of Chicago
Sadhu Johnston
Rick Valicenti and John Pobojewski, Thirst